SpringerBriefs in Optimization

Series Editors

Panos M. Pardalos
János D. Pintér
Stephen M. Robinson
Tamás Terlaky
My T. Thai

SpringerBriefs in Optimization showcases algorithmic and theoretical techniques, case studies, and applications within the broad-based field of optimization. Manuscripts related to the ever-growing applications of optimization in applied mathematics, engineering, medicine, economics, and other applied sciences are encouraged.

For further volumes:
http://www.springer.com/series/8918

Elisa Pappalardo • Panos M. Pardalos
Giovanni Stracquadanio

Optimization Approaches for Solving String Selection Problems

Springer

Elisa Pappalardo
Department of Biomedical Engineering
Johns Hopkins University
Baltimore, MD, USA

Giovanni Stracquadanio
Department of Biomedical Engineering
Johns Hopkins University
Baltimore, MD, USA

Panos M. Pardalos
Center for Applied Optimization
Department of Industrial and Systems
   Engineering
University of Florida
Gainesville, FL, USA

Laboratory of Algorithms and Technologies
   for Networks Analysis (LATNA)
Higher School of Economics
National Research University
Moscow, Russia

ISSN 2190-8354   ISSN 2191-575X (electronic)
ISBN 978-1-4614-9052-4   ISBN 978-1-4614-9053-1 (eBook)
DOI 10.1007/978-1-4614-9053-1
Springer New York Heidelberg Dordrecht London

Library of Congress Control Number: 2013949158

Mathematics Subject Classification: 90C05, 90C10, 90C11, 90C27, 90C56, 90C90, 92C50, 92B05, 97M60

Printed on acid-free paper

Springer is part of Springer Science+Business Media (www.springer.com)

Preface

This monograph discusses some of the most well-known string selection problems (SSP), offering a compendium of the current state-of-the-art methods presented in the literature. In particular, this work is intended as a general and comprehensive guide to understanding the very basic notions of mathematical optimization for sequence selection in biology, and a review of current research directions in this area; it aims to help bridge the gap between the computational and biological aspects of this class of problems.

SSP are generally modeled as optimization problems addressing the selection of strings with specific features from a large set of input sequences. In this context, the notion of "specific features" is captured by an objective function and we want to find strings that minimize or maximize it. Many problems in computational biology address the task of determining meaningful properties of biological systems, by comparing sequences, discovering their similarities and/or differences. From probe and primer design for diagnostics to protein structure prediction, from disease modeling to drug design, many steps of such design processes are characterized by discovering similarities or differences among sequences. However, one of the main challenges that arise when studying optimization methods for biological applications is the need of finding meaningful and high quality results, from both computational and biological points of view. Though theoretical results are usually desirable from a mathematical point of view, practical implications are more attractive from a biological perspective; this is a well-known issue in computational biology, where working at the edge of mathematics and biological sciences usually requires to find a trade-off between the biological findings and the in silico results.

Preliminarily, in Chap. 1 we provide a brief introduction to some biological concepts to understand the scenario where SSP arise; in Chap. 2 we introduce the basic notations for representing biological sequences, such as the Levenshtein distance and the Hamming distance, and we give an overview of biological applications that require the solution of SSP. Successively, we introduce some basic notions of mathematical optimization in Chap. 3, aiming at providing a minimal background to non-experts, in order to understand the basis of string selection methods, which are covered in Chap. 4. We introduce some of the most studied

SSP in computational biology and discuss some of the existing approaches, with an emphasis on their computational complexity. As such problems are mainly NP-hard, we put the focus on methods with theoretical guarantee on the quality of solutions and on methods with no theoretical proof but fast and effective in practice. In this perspective, several approaches can be considered, ranging from exact to heuristic methods; although exact methods guarantee to find a locally optimal solution, their computational burden makes them impractical when dealing with large datasets. Conversely, approximated methods guarantee a solution with a bounded distance from the optimum, while heuristics simply seek for a good solution without making any assumption on its optimality. Nevertheless, the latter two approaches allow to overcome the high computational cost required by exact methods and usually reach good performance even for large datasets.

At the end of our discussion on the state-of-the-art methods for string selection, and the comparison, when possible, of the approaches proposed in literature, we draw out the conclusions of this monograph. Many questions remain to be answered, both from theoretical and practical points of view, and many topics of interest remain open for further investigation.

Acknowledgments

Panos M. Pardalos was partially supported by LATNA Laboratory, NRU HSE, RF government grant, ag. 11.G34.31.0057

Baltimore, MD   Elisa Pappalardo

Contents

Chapter 1
Biological Sequences

Abstract Starting with the introduction of the first chain-termination method for DNA sequencing in 1977 by Sanger *et al.*, the ability of unveiling the genome of an organism marked a breakthrough moment in biology. Nowadays, the deluge of data provided by high-throughput sequencing technologies continuously requires the design of fast and efficient analysis methods. Moreover, the progress of DNA synthesis technologies requires algorithms to design molecules with specific functions. In this chapter, we introduce some basic biological concepts and how DNA sequences are obtained from an organism through sequencing, to set the fundamentals to understand the importance of string selection algorithms.

1.1 Introduction

The first draft of the euchromatic sequence of the human genome in 2001 by the *International Human Genome Sequencing Consortium* [4] and *Celera Genomics* [10] revolutionized the field of biology, providing an extraordinary source of information on the structure and function of a human cell. Determining the sequence of an organism is one of the most challenging interdisciplinary scientific problems; in particular, the impressive amount of data generated by the sequencing process requires computational methods that can provide high-quality results in a reasonable amount of time.

The introduction of the pyrosequencing method in 2005 [5] represented a breakthrough in genomics; this new technology, largely known as next-generation sequencing, allowed to sequence thousands of DNA molecules in parallel producing a 100 fold increase in throughput compared to Sanger sequencing and leading to hundreds of thousands bases sequenced per run. Nowadays, the introduction of single-molecule real time sequencing [3] and semiconductor sequencing [7] is leading towards the era of personal genome sequencing, with a whole human genome sequencing closely approaching the cost of $1,000. The impact of high-throughput

E. Pappalardo et al., *Optimization Approaches for Solving String Selection Problems*,
SpringerBriefs in Optimization, DOI 10.1007/978-1-4614-9053-1_1,
© Elisa Pappalardo, Panos M. Pardalos, Giovanni Stracquadanio 2013

sequencing is not limited to unveil the genome of an organism, but also to understand the transcriptome structure and gene expression [6].

In computational terms, sequencing is the method used to decode the information encoded inside an organism; understanding what these information represent and how the decoding process works is fundamental to design effective and efficient algorithms. First, we introduce some basic notions about DNA, RNA and proteins and, successively, we will give an overview of some widely used sequencing technologies.

1.2   DNA, RNA and Proteins

The flow information in a cell is coherent with the *central dogma of molecular biology*: "DNA is transcribed into RNA which is translated into Proteins" [2].

The deoxyribonucleic acid (DNA) is a polymer that carries the genetic information in the cell; its three-dimensional structure consists of two antiparallel complementary strands of monomers, called nucleotides. Each nucleotide comprises three components: a 2-deoxyribose, a phosphate group and a nucleobase. Nucleobase and sugar form the nucleoside. Nucleobases containing nitrogen are classified as purines or pyrimidines: the purines, adenine (A) and guanine (G), are characterized by a six-membered nitrogen-containing ring fused to a five-membered ring. The pyrimidines have only a six-membered nitrogen-containing ring, and are cytosine (C) and thymine (T). The two strands of DNA are structured in such a way that an adenine on one strand is attached to a thymine on the other strand, and the guanine binds to cytosine. This binding mechanism is known as Watson-Crick DNA complementary base pairing [1].

Although only four different nucleotides constitute the building block of DNA, each nucleic acid contains millions of them, and the order in which they appear codes the information carried by the nucleic acid. Specifically, the exact sequence of nucleotides on a DNA strand forms the genetic code, and the entire genetic code of an organism represents its genome. A gene is a functional unit of the genome, which encodes the information for making proteins; in general, a gene is characterized by a coding region, which encodes the aminoacid sequence of a protein, and a regulatory region that is responsible for starting and stopping protein production.

Gene information is converted into proteins in two steps, called *transcription* and *translation*. During transcription, the coding region of a gene is transcribed into a single stranded ribonucleic acid (RNA); like DNA, RNA has a long chain of nucleotide units, consisting of a nitrogenous base, a ribose sugar, and a phosphate. While DNA contains deoxyribose, RNA bases contain ribose, and the base uracil (U) rather than the thymine. Because A, G, C, and T appear in DNA, these molecules are called DNA-bases, whereas A, G, C, and U are called RNA-bases. During transcription, DNA is used as a template to synthesize messenger RNA (mRNA). Transcription relies on the principle of complementarity base pairing: an enzyme, the RNA polymerase, reads one of the two strands of a DNA molecule, and assembles bases that are complementary to the DNA strand to make an RNA transcript [1].

After transcribing the coding region, RNA may be further edited before its translation. In prokaryotic cells the transcription mechanism generates mRNA, which does not need any post-transcriptional modification; conversely, in eukaryotic cells, the transcript must be further modified before it can be used. Post-transcriptional changes include the addition of more adenine bases at the tail region, removal of introns, i.e. regions that do not code for protein synthesis, and exons splicing. Exons represent the coding regions, which are joined together in this phase to form a single chain, representing the mature mRNA transcript; this mechanism is extremely important and allows a gene to encode for multiple proteins [1].

RNA transcripts are successively carried to the cytoplasm of the cell where the translation process occurs. Translation involves three kinds of RNA molecules: messenger RNA (mRNA), ribosomal RNA (rRNA), and transfer RNA (tRNA). In the cytoplasm, mRNA is attached to a cellular structure, called ribosome; mRNA bases are read three at a time (codon) and translated into an amino acid to be incorporated into the protein being synthesized [1].

Amino acids are monomers characterized by a central α carbon atom (C_α) bonded to four different chemical groups; an amino (NH2) group, a carboxyl (COOH) group, a hydrogen (H) atom, and a side chain, which is amino acid specific. There are 20 amino acids encoding for a plethora of proteins. Amino acids do not interact directly with mRNA in protein synthesis, but they are carried to the messenger RNA by the tRNA, a mechanism controlled by a ribosome. Ribosomes serve as to carry and host the enzymes necessary for protein synthesis, and consist of various proteins plus the ribosomal RNA, that links amino acids together to form proteins. For accomplishing protein synthesis, the tRNA uses a set of three nucleotide bases at one end that are complementary to a corresponding codon; the tRNA triplets are called anticodon. The tRNA reads the first mRNA codon by using its own anticodon; this represents a critical step since anti codons use the information coded in mRNA to determine the amino acid to be added to the chain. Different codons can encode the same amino acid, leading to what is known as *degeneracy of the genetic code* [1].

1.3 Sequencing Methods

Determining the sequence of nucleotides forming a genome is possible by sequencing DNA molecules. Currently, there are two main approaches for DNA sequencing; the chain-terminating dideoxynucleotides method, knows as *Sanger* sequencing [8], and high-throughput parallel sequencing methods, called *next-generation* methods.

Sanger sequencing requires a single-stranded DNA template, DNA polymerase, deoxynucleosidetriphosphates (dNTPs) and special nucleotides (dideoxyNTPs or ddNTPs), which lack the 3'-OH group causing the termination of the elongation of DNA strands. In its basic form, there are four independent sequencing reactions, one for each standard dideoxynucleotides; each reaction contains also dNTPs and DNA

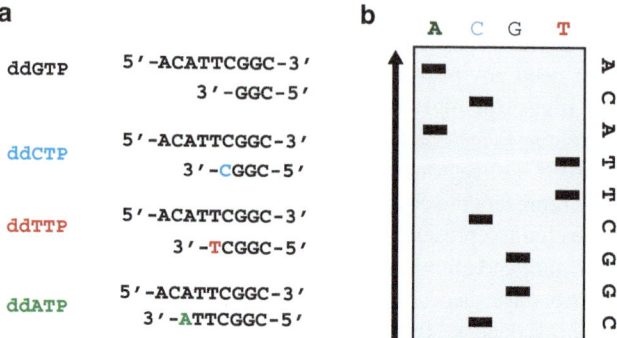

Fig. 1.1 Sanger sequencing. (a) ddNTPs incorporation at different point of template DNA sequence. (b) Gel electrophoresis of chain-terminated DNA molecules

polymerase. The polymerase incorporates dNTPs and ddNTPs at random, causing the strand to be cut short when a ddNTP is picked; by running the population of chain-terminated DNA molecules side by side on a high resolution gel, the bands appearing on the gel tell which nucleotide is at each position by reading from the top to the bottom (see Fig. 1.1). However, the most widely used Sanger sequencing approach is to run a single reaction with all the four ddNTPs, which are fluorescently labelled with different colors [9]; the chain-terminated fragments are sorted through capillary electrophoresis machines and a laser reads the color of the fragments in order, producing in output a chromatogram, which provides the nucleotide sequence and the fluorescence peaks for each base. Peaks are converted to the Phred quality scores, which assess the probability of a nucleotide of being incorrectly identified.

The increasing need of DNA sequencing at a genome scale makes impractical the adoption of Sanger sequencing; moreover, with the genomic research focusing on the discovery of variations among different individuals, there is the demand of generating sequencing data in high-throughput fashion. Next-generation sequencing methods rely on a completely different approach respect to the Sanger method; when the polymerase incorporates a nucleotide, the pyrophosphate release is converted into Adenosine-triphosphate (ATP) by the sulfurylase enzyme, producing firefly luciferase with the necessary energy to convert luciferin to oxy-luciferin and light. Each template is queried with a single nucleotide and, hence, the emitted light from each reaction can be apportioned to the incorporation of a certain number of nucleotides [5]. This sequencing process is fairly easy to parallelize, with several commercial implementations available on the market; nowadays, the most commonly used platforms are provided by Illumina® (Fig. 1.2).

First, a genomic library is prepared by fragmenting the genomic DNA to a predefined length (e.g. 500bp) and end repaired by ligating platform specific adapters. Successively, fragments are hybridized to a complimentary oligo that is linked to a glass pane, called *flow cell*, and amplified by Polymerase Chain Reaction (PCR); amplified fragments are denatured and annealed with a sequencing

Fig. 1.2 Illumina sequencing by synthesis. (a) Genomic DNA is obtained and sheared into fragments of predefined length. (b) Adapters are ligated to the fragments. (c) Fragments are hybridized to complimentary oligos on the flow cell and bridge amplified. (d) Double stranded bridges are denatured to obtain a single strand DNA molecule with only one end attached to the surface. (e) Four fluorescent terminators (e.g. A=green, C=blue, T=red, G=black) are added together with primers and DNA polymerase and, after laser excitation, the emitted fluorescence is captured and imaged. (f) The process is iterated for several cycles

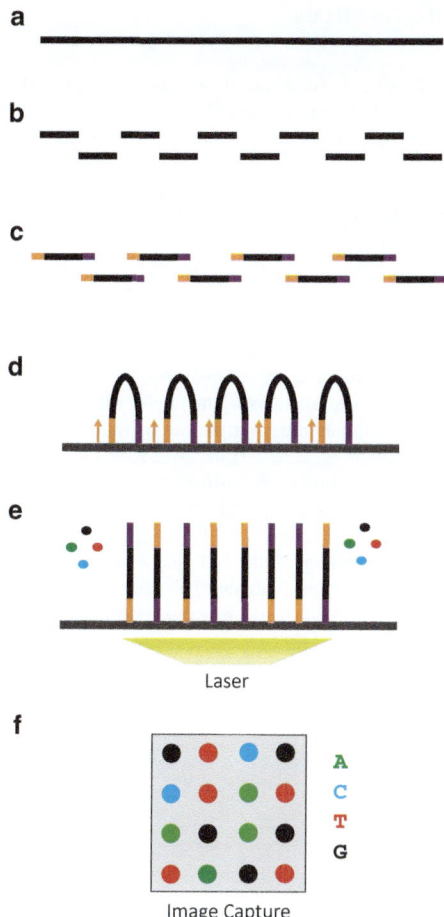

primer and deployed to the sequencing machine for sequencing-by-synthesis using fluorescent dNTPs. At each sequencing cycle, a single nucleotide is incorporated, and the flow cell washed to remove the unincorporated nucleotides; after the whole flow cell is imaged, the 3'-OH ends on the synthesized strands are removed and the flow cell is washed again. Illumina machine are able to generate 120Gb in about 27 hours or 600Gb in about 11 days using Illumina® HiSeq 2500.[1]

---

[1]http://www.illumina.com/systems/sequencing.ilmn

References

1. Alberts, B., Johnson, A., Lewis, J., Raff, M., Roberts, K., Walter, P.: Molecular biology of the cell, 4 edn. Garland (2002)
2. Crick, F., et al.: Central dogma of molecular biology. Nature **227**(5258), 561–563 (1970)
3. Eid, J., Fehr, A., Gray, J., Luong, K., Lyle, J., Otto, G., Peluso, P., Rank, D., Baybayan, P., Bettman, B., et al.: Real-time dna sequencing from single polymerase molecules. Science **323**(5910), 133–138 (2009)
4. Lander, E.S., Linton, L.M., Birren, B., Nusbaum, C., Zody, M.C., Baldwin, J., Devon, K., Dewar, K., Doyle, M., FitzHugh, W., et al.: Initial sequencing and analysis of the human genome. Nature **409**(6822), 860–921 (2001)
5. Margulies, M., Egholm, M., Altman, W.E., Attiya, S., Bader, J.S., Bemben, L.A., Berka, J., Braverman, M.S., Chen, Y.J., Chen, Z., et al.: Genome sequencing in microfabricated high-density picolitre reactors. Nature **437**(7057), 376–380 (2005)
6. Mortazavi, A., Williams, B.A., McCue, K., Schaeffer, L., Wold, B.: Mapping and quantifying mammalian transcriptomes by rna-seq. Nature methods **5**(7), 621–628 (2008)
7. Rothberg, J.M., Hinz, W., Rearick, T.M., Schultz, J., Mileski, W., Davey, M., Leamon, J.H., Johnson, K., Milgrew, M.J., Edwards, M., et al.: An integrated semiconductor device enabling non-optical genome sequencing. Nature **475**(7356), 348–352 (2011)
8. Sanger, F., Nicklen, S., Coulson, A.R.: Dna sequencing with chain-terminating inhibitors. Proceedings of the National Academy of Sciences **74**(12), 5463–5467 (1977)
9. Smith, L.M., Sanders, J.Z., Kaiser, R.J., Hughes, P., Dodd, C., Connell, C.R., Heiner, C., Kent, S.B., Hood, L.E.: Fluorescence detection in automated dna sequence analysis. Nature **321**, 674–679 (1986)
10. Venter, J.C., Adams, M.D., Myers, E.W., Li, P.W., Mural, R.J., Sutton, G.G., Smith, H.O., Yandell, M., Evans, C.A., Holt, R.A., et al.: The sequence of the human genome. Science Signaling **291**(5507), 1304–1351 (2001)

Chapter 2
Strings: Theory, Properties and Applications

Abstract Genomic and proteomic data can be represented as sequences over the nucleotides and amino acids alphabets, and many tasks require algorithms working on strings. This chapter introduces the formalism to deal with sequences and the definition of distance metrics, largely used in string selection methods.

2.1 Introduction

Sequencing projects generate a deluge of data, which is represented by sequences over an alphabet Σ (e.g., the nucleotides alphabet); in this context, string selection is a fundamental task that finds application in different fields, e.g. phylogenetic tree reconstruction, primer design, DNA binding sites identification. Moreover, the ability to synthesize long DNA molecules has increased the need of algorithms to identify sequences that meet specific requirements, e.g. minimize homopolymer segments, design of optimal assembly oligos. Nevertheless, many of these problems are NP-hard, representing challenging tasks both from a computational and biological perspectives. In this chapter, we introduce the basic notations for representing biological sequences; since many tasks involve finding similarities and common patterns, we introduce the most largely used metrics, which are the Levenshtein distance and the Hamming distance. Finally, we give an overview of biological applications that require the solution of string selection problems.

2.2 Basic Definitions

DNA, RNA, and protein sequences can be thought as strings of symbols over a finite alphabet. Specifically, the DNA alphabet consists of four characters, corresponding to the nucleotides, while protein sequences can be viewed as strings over the

E. Pappalardo et al., *Optimization Approaches for Solving String Selection Problems*, SpringerBriefs in Optimization, DOI 10.1007/978-1-4614-9053-1__2,

7

20-letter alphabet of amino acids. In this section, we introduce some fundamental concepts, in order to settle the notation for this manuscript.

An alphabet $\Sigma = \{c_1, \ldots, c_k\}$ is a finite set of elements, called characters. A string s can be defined as a finite sequence of characters $(c_1, \ldots, c_m)$, $c_i \in \Sigma$, $i = 1, \ldots, m$, and we denote the empty string $s^0 = \epsilon$. Given a string s over a finite alphabet Σ, $|s|$ and s_i denote the length of s and the ith character of s, respectively.

2.3   Distance Metrics

Discovering analogies and/or differences in genomic data requires the introduction of metrics to quantify the similarity, alternatively the distance, between two sequences. In general, two metrics are commonly adopted: the Levenshtein or edit distance [5] and the Hamming distance [3].

The Levenshtein edit distance between two strings s and t, denoted as $d_L(s, t)$, is the minimum number of edits, which are insertions, deletions, and substitutions, required to transform one string into the other. Let us define the predicate function $\Phi_L(s, t)$ between $s_1, \ldots, s_i$ and $t_1, \ldots, t_j$ as follows:

$$
\Phi_L(i, j) = \begin{cases} 0 & \text{if } i = j = 1 \\ i & \text{if } j = 1 \wedge i > 1 \\ j & \text{if } i = 1 \wedge j > 1 \\ \min \begin{cases} \Phi_L(i-1, j) + 1 \\ \Phi_L(i, j-1) + 1 \\ \Phi_L(i-1, j-1) + 1 + [s_i \overset{?}{=} t_j] \end{cases} & \text{otherwise} \end{cases} \tag{2.1}
$$

where $s_i \overset{?}{=} t_j$ is equal to 1 if the characters s_i and t_j match, otherwise is 0.

The first element in the minimum corresponds to an insertion from s to t, the second to a deletion and the third to a match or a mismatch, depending on whether the respective symbols are the same. The Levenshtein distance for the string s and t is defined as

$$
d_L(s, t) = \Phi_L(|s|, |t|). \tag{2.2}
$$

Since the various type of edit operations occur with different probabilities, a metric which takes into account weights can be more reliable: in biology, for instance, some mismatches are less penalizing than others (e.g., mismatch involving amino acids of the same family). In this case, we define the weighted edit distance between two strings s and t as the cost of the cheapest sequence of edit operations needed to transform s into t, where weights are considered.

The Hamming distance represents a special case of the edit distance, where only mismatches between strings are taken into account. Specifically, the Hamming distance between two strings s and t having equal length, denoted by $d_H(s, t)$, is the number of positions at which s and t differ. Let $\Phi_H : \Sigma \times \Sigma \rightarrow \{0, 1\}$ a predicate function such that $\Phi_H(x, y) = 1$ if and only if $x \neq y$; the Hamming distance between two strings s and t is defined as:

$$d_H(s, t) = \sum_{i=1,\ldots,|s|} \Phi_H(s_i, t_i). \qquad (2.3)$$

Although the edit distance is a more accurate metric for several genomic operations, many problems are defined in terms of Hamming distance. There are three main reasons: first, the problems we analyze have the objective of finding an exact string or substring, not a modified version of it, therefore gaps are not taken into account. Moreover, since gaps are more destabilizing than substitutions, the use of the Hamming distance is preferable as a distance metric [4]. Last, the Hamming distance can be used to describe the effects of mutations in the protein coding regions of DNA, whereas edit distance is more suitable to measure the evolutionary distance [4].

2.4 Applications

The objective of our analysis are problems addressing the localization of similar features in nucleotide or amino acids sequences, and the identification of patterns enriched in a set of sequences. The first class of problems comprises the closest string problem (CSP), the closest substring problem (CSSP) and its decision version, the common approximate substring problem (CAS), the close to most string problem (CMSP), the center and median string problems; the second class includes the farthest string problem (FSP), the farthest substring problem (FSSP), and the far from most string problem (FFMSP). Another class of problems mentioned in this work aims at finding a pattern that occurs in one set of strings but does not occur in another set, known as distinguishing string selection problem (DSSP), and the d-mismatch problem, which generalizes the concept of closest string to center strings of an aligned set of substrings. These problems arise in many molecular biology tasks and, hence, finding high-quality solutions is challenging both for computer scientists and for biologists; below, we will present some molecular biology problems strictly related to SSP.

2.4.1 Primer Design for Polymerase Chain Reaction

Polymerase chain reaction (PCR) is a technique adopted in molecular biology for amplifying a portion of DNA in many copies. PCR has many applications, such as DNA cloning for sequencing, functional analysis of genes, forensic, disease diagnosis. First, PCR requires the selection of two primers, which are fragments of DNA complementary to the $3'$ ends of the sense and antisense strands of the regions to amplify, called template DNA; the primers bind the template DNA during the annealing step of PCR, and the polymerase binds to these regions to start DNA synthesis.

Primer selection is a complex task, and it affects the results of PCR experiments. Particularly interesting is the selection of primers that are able to amplify several regions simultaneously; recently, it has been shown that this task can be addressed as a DSSP [1].

2.4.2 Identification of Transcription Factor Binding Sites

A phylogenetic tree or evolutionary tree is a diagram that depicts the evolutionary relationships among various species or other entities, based upon similarities and differences in their characteristics. Phylogenetic trees can be inferred from sequence alignment and used to detect potentially important regions within a DNA sequence: given an alignment, we look for highly conserved regions.

Transcription factors are proteins that bind to specific DNA sequences, controlling the transcription of genetic information from DNA to mRNA, thus affecting the expression of a gene. In particular, transcription factors bind to a specific DNA site, allowing only a small amount of variation; therefore, they can be used to identify conserved regions in biological sequences. Identifying transcription factors among a set of sequences can be reduced to the d-mismatch problem [8]; given a set of sequences, possible binding sites can be identified as a string matching the input sequences with at most d mismatches.

2.4.3 Multiple Tree Alignment Problem

A multiple alignment is a sequence alignment of three or more biological sequences, such as proteins, DNA, or RNA, used to identify regions of similarity that may imply functional, structural, or evolutionary relationships among the sequences. The problem can be viewed as finding a set of patterns which, with some errors, appear in the same order in all the sequences of a given set [2]. A variant of the problem is known as the tree alignment with a given phylogeny; given a set S of n sequences over an alphabet Σ, and an unlabelled tree with n leaves, representing the phylogeny, an evolutionary tree is a labelled tree built on the phylogeny, having labels on the leaves representing the sequences in the input set S, and labels on the other nodes which are the sequences over the alphabet Σ.

The cost of an evolutionary tree is the sum of all the edit distances between the labels of pairs of nodes joined by a vertex, and an optimal evolutionary tree is a tree that minimizes such cost [2]. A star phylogeny is characterized by $n + 1$ nodes, where n of them are leaves of the tree. This topology is often used to represent a recent population expansion event from a common ancestor. Such problem can be viewed as finding the median string of a set of input sequences.

2.4.4 Design of Diagnostic Probes

The task of designing a string that is able to represent a set of known sequences and, at the same time, is easily distinguishable from another set, is a problem arising in the diagnosis of viruses and bacteria in host organisms. In particular, probes are used to diagnose the presence of viruses and bacteria in biological samples. A probe is a strand of DNA or RNA opportunely treated with a radioactive isotope, dye, or enzyme, and used to easily detect the presence of a specific sequence of nucleotides, called target, on a single-stranded nucleic acid, by performing hybridization experiment. The probe hybridizes to the target if they are complementary to each other. Hence, given a set of sequences representing the virus or the bacteria, and a host, the problem is to discover a sequence that occurs in virus sequences, but does not appear in the host [7]. This problem involves the design of the probe sequence, that has to be as close as possible to the sequences to detect, and, on the other hand, as far as possible from another set of sequences.

2.4.5 Protein Function Prediction

Predicting the function of an unknown protein represents a central problem in bioinformatics, hence an increasing number of methods have been proposed in literature to tackle this problem [6]. One approach consists in analyzing the shared motifs, which are common motifs between two or more sequences and can be associated with a particular function or regulation. String selection problems find application in addressing such problems, by looking for suboptimal matchings among strings.

2.4.6 Drug Design

Another important application of string selection concerns the design of drugs and therapies. Here, given a set of sequences of orthologous genes from a group of closely related pathogens, and a host, the goal is to identify a sequence that is highly conserved in all or most of the pathogens' sequences, but that is not present in the host [4]. In fact, conserved regions might encode relevant biological information, since they seem resilient to mutations. This information can be exploited to identify the chemical components that bind the conserved region, in order to create new effective therapies. In antisense drug design, the same principle is used to create drugs that inhibit the production of the protein related to the disease, but do not interfere with other proteins [4]. Specifically, antisense therapies focus on impeding the production of proteins that cause the disease: antisense drugs bind mRNA to prevent the genetic code related to the disease be read by the ribosome, which is responsible in assembling proteins based on the instructions carried by mRNA.

References

1. Boucher, C.: Combinatorial and probabilistic approaches to motif recognition. Ph.D. thesis, University of Waterloo (2010)
2. de la Higuera, C., Casacuberta, F.: Topology of strings: Median string is NP-complete. Theor. Comput. Sci. **230**(1), 39–48 (2000)
3. Hamming, R.W.: Error detecting and error correcting codes. Bell Syst. Tech. J. **29**(2), 147–160 (1950)
4. Lanctot, J., Li, M., Ma, B., Wang, S., Zhang, L.: Distinguishing string selection problems. In: Proceedings of the Tenth Annual ACM-SIAM Symposium on Discrete Algorithms, pp. 633–642 (1999)
5. Levenshtein, V.I.: Binary codes capable of correcting deletions, insertions and reversals. In: Soviet Physics Doklady, vol. 10, p. 707 (1966)
6. Marcotte, E.M., Pellegrini, M., Ng, H.L., Rice, D.W., Yeates, T.O., Eisenberg, D.: Detecting protein function and protein-protein interactions from genome sequences. Science **285**(5428), 751–753 (1999)
7. Phillippy, A.M., Mason, J.A., Ayanbule, K., Sommer, D.D., Taviani, E., Huq, A., Colwell, R.R., Knight, I.T., Salzberg, S.L.: Comprehensive dna signature discovery and validation. PLoS Comput. Biol. **3**(5), e98 (2007)
8. Stojanovic, N., Berman, P., Gumucio, D., Hardison, R., Miller, W.: A linear-time algorithm for the 1-mismatch problem. In: Algorithms and Data Structures, pp. 126–135. Springer, Berlin Heidelberg (1997)

Chapter 3
Mathematical Optimization

Abstract Mathematical Optimization is an interdisciplinary branch of applied mathematics, related to the fields of Operations Research, Computational Complexity, and Algorithm Theory. It can be informally defined as the science of finding the "best" solution from a set of available alternatives, where the notion of "best" is intrinsically related to the specific problem addressed. Nowadays, optimization problems arise in all sorts of areas and are countless in everyday life, such as in engineering, microelectronics, telecommunications, biomedicine, genetics, proteomics, economics, finance, and physics. However, in spite of the proliferation of optimization algorithms, there is no universal method suitable for all optimization problems, and the choice of the "most appropriate method" to solve the specific problem is demanded to the user. With this in mind, in this chapter we address some general questions about optimization problems and their solutions, in order to provide a background knowledge of the issues analyzed later.

3.1 Preliminaries

Three major problem domains can be defined in computer aided design: *modeling*, *simulation*, and *optimization*. The purpose of modeling and simulation is to design and tune models that capture the essential behavior of systems and processes, typically, by introducing a set of algebraic and differential equations. In this context, optimization aims at finding an optimal solution for a set of properties of a system, under given constraints.

Optimization is a crucial task in engineering systems, and arises in every area of science, ranging from industry-related problems, such as supply chain management or resources allocation, to biology, like protein structure prediction or metabolic engineering [26]. In its simplest form, the optimization process seeks to minimize or maximize the value of a function by finding the values of the variables from a given domain. However, for many problems, it is not known whether and how they can be solved exactly in polynomial time [23, 24]; when the problem cannot be

E. Pappalardo et al., *Optimization Approaches for Solving String Selection Problems*,
SpringerBriefs in Optimization, DOI 10.1007/978-1-4614-9053-1_3,
© Elisa Pappalardo, Panos M. Pardalos, Giovanni Stracquadanio 2013

solved in polynomial time, the requirements are relaxed towards the identification of *near-optimal* solutions, where the notion of *optimal* is strictly problem specific.

In this chapter, we introduce the basic concepts and notations used in mathematical optimization, by pointing out both the theoretical and practical aspects; after a formal definition of optimization problems, we focus our attention on the complexity of the problems and the quality of solutions. Successively, we discuss different classes of optimization problems and methods.

3.2   Optimization Problems

Mathematical Optimization is the process of formulating and solving an optimization problem, possibly subject to a series of constraints; in particular, the optimization process aims at minimizing or maximize the value of a function, called *objective function*, by choosing integer or real variables from a predefined domain. Without loss of generality, we will restrict our discussion to minimization, although same considerations hold for maximization problems.

A standard form for an optimization problem can be defined as follows [3]:

$$\begin{aligned}
\min \; & f(x) \\
& g_i(x) \leq 0, \; i = 1, \ldots, m \\
& h_j(x) = 0, \; j = 1, \ldots, k
\end{aligned} \tag{3.1}$$

where $x = (x_1, x_2, \ldots, x_n)$ is the vector of decision variables, $f : \mathbb{R}^n \to \mathbb{R}$ is the *objective function*, $g_i : \mathbb{R}^n \to \mathbb{R}, i = 1, \ldots, m$ represent inequality constraints, whereas $h_j : \mathbb{R}^n \to \mathbb{R}, j = 1, \ldots, k$ the equality constraints.

The objective function is the quantity we want to minimize, which depends on the decision variables, namely it has to be computable from the values of such variables. A constraint is a restriction that any valid solution must satisfy; only a solution that satisfies all of these constraints, called *feasible* solution, is acceptable [27].

A vector x^* is called *optimal*, or a solution of the problem (3.1), if it has the smallest objective function value among all vectors that satisfy the constraints, that is for any feasible solution z, $f(z) \geq f(x^*)$.

As an illustrative example of an optimization problem, we consider the "diet problem," one of the first modern optimization problems studied [4]; the goal is to find the cheapest combination of foods so that the daily nutritional requirements of a person are satisfied. The objective function aims at minimizing the cost of the foods purchased, the decision variables are the amounts of each type of food to purchase (supposed as continuous values). Since we require that the purchased foods provide at least a certain amount of each nutrient, the constraints are the nutritional needs to be satisfied, such as proteins, vitamins, minerals, and calories. Thus among all the possible adequate meals, the problem is to find one that is least costly. This problem presents some specific features, according to whom some classifications can be made. In Sect. 3.5, we introduce and describe the major classes of optimization problems.

3.3 Computational Complexity

Despite the fact that many real-world problems can be modeled as optimization problems, they are often very difficult to solve. In order to study the tractability of a problem, we need a framework to quantify the resources needed to find a feasible solution. Choosing quantitative measures, such as run-time or memory requirements, is irrelevant, since they depend on the computing platform and on the specific algorithm implementation; moreover, the time required to solve an instance of a problem depends on certain properties of the instance itself. In general, we are interested to study the computational effort required to find a feasible solution as a function of the size of the input.

Computational complexity analyzes the inherent intractability or tractability of computational problems and classifies them into complexity classes according to their computational hardness [8]. Specifically, if we consider computational time, two main classes can be identified, which are P and NP. The first complexity class contains all the problems that can be solved by polynomial-time algorithms, and here P stands for polynomial time. P can be considered as the class of tractable problems that can be solved efficiently. Conversely, NP indicates the class of nondeterministic polynomial problems, that is the set of decision problems solvable in polynomial time by a non-deterministic Turing machine. It is not known whether "P = NP", though it is widely believed that there are problems in NP which are not in P. Some such problems have a property known as NP-*completeness*. NP-complete problems are considered intractable, and only approximate solutions can be found. In particular, if an NP-complete problem could be solved in polynomial time, thus all NP-complete problems can be solved in polynomial time, with the consequence that P = NP [8].

Formally, NP-completeness is defined as follows:

Definition 1 (NP-Completeness). A decision problem C is NP-*complete* if C is in NP, and every problem in NP is reducible to C in polynomial time.

Specifically, a problem A is said to be "reducible in polynomial time" to a problem B if, for each instance a of A, it is possible to build an instance b of B in polynomial time, such that the two instances have the same truth values. A problem satisfying the second condition of NP-*completeness*, but not necessarily the first one, is said to be NP-*hard*. Informally, a NP-hard problem is "at least as difficult as" every other NP-complete problem [8].

Although NP-complete problems are generally thought of as being computationally intractable, for some given problems hard instances are rare, thus they can be solved efficiently in practice. In order to measure how a problem behaves "in general" and in the worst case, the average-case and the worst-case complexity can be analyzed. Shortly, worst-case and average-case complexity measure the resources an algorithm requires in the worst-case scenario and on average, such as running-time and memory. Worst-case complexity gives an upper bound of the computational efforts needed.

3.4   Global and Local Optimal Solutions

In the previous sections, we generically referred to optimal solutions of an optimization problem; such notation adheres to the definition of "global optimum." A *globally optimal solution* for an optimization problem is defined as the solution $x^* \in \mathbb{R}^n$, such that $f(x^*) \leq f(x)$, for all $x \in \mathbb{R}^n$. In order to define a global optimum, therefore, it is not necessary to introduce the structure of the search space, or a neighborhood. However, in general, finding the global optimum of an optimization problem can be a very difficult task, and it is computationally intractable for many problems [13].

Nevertheless, it is often possible to find a solution that is the best within a specific subset of the space of solutions, called *neighborhood*: this solution represents a *local optimum* for the problem. Given a feasible point $x \in \mathbb{R}^n$, a *neighborhood* can be informally defined as a set $N \subseteq \mathbb{R}^n$ of points that are close, in some sense, to the point x, e.g. they are computed starting from x, or that share a significant part of their structure with x.

The search of a local optimum is a task addressed by local search methods. In general, this class of methods starts from an initial solution and iteratively modifies it into a better one, by performing a move on it, i.e. a small perturbation; the move identifies a neighbor solution. We say that a solution $\hat{x} \in \mathbb{R}^n$ is locally optimal with respect to the neighborhood N, or N-optimal, if for all $x \in N(\hat{x})$, $f(\hat{x}) \leq f(x)$.

Two important issues in local search methods are the quality of the solution found, and the complexity of the local search heuristic. Usually, the best choice is a trade-off between quality and computational complexity: whereas the larger is the neighborhood, the better will potentially be the solution found, it might be computationally intractable to compute it. Therefore, the design of a local search heuristic involves the choice of a promising neighborhood, that is a neighborhood large enough to contain good solutions, and on the other hand, small enough to ensure an efficient search. In general, the neighborhood function may, or may not, generate the globally optimal solution. When the neighborhood function is able to generate the global optimum, starting from any initial feasible point, it is called *exact*.

3.5   Classification

In Sect. 3.2, we have introduced the "diet problem"; such problem has specific properties: it is a continuous problem, and both the objective function and the constraints are linear with respect to the decision variables. Therefore, the diet problem belongs to the class of linear programming problems. We generally consider classes of optimization problems, defined by specific features of the objective function or constraints. In the following section, we shortly introduce the major families, according to some specific classification criteria.

3.5.1 Combinatorial Versus Continuous Optimization Problem

First, we analyze optimization problems with respect to the type of the variables involved; according to this criterion, problems can be divided into two main categories: those involving *continuous* variables, and those with *discrete* variables [6].

Combinatorial optimization represents the subset of mathematical optimization where the solution domain is discrete or can be reduced to discrete. In general, the class of combinatorial optimization problems is considered more difficult than the continuous optimization problem set. A main difference is represented by the notion of neighborhood: while, for the first class of problems, the neighborhood of a solution is defined by a small change in the combination of the elements of a solution, in continuous optimization small movements and search directions are taken into account [20]. In continuous optimization we can make use of gradient information, unlike discrete problems, where the lack of direction makes this approach impracticable. Given a function f, the gradient for a point x_i is defined as:

$$\frac{\partial f}{\partial x_i} = \frac{f(x_i + \varepsilon) - f(x_i)}{\varepsilon} \tag{3.2}$$

which informally represents the direction of the greatest decrease of f starting from x_i, thus it can be used to select the most promising neighborhood to identify a promising region of the search space. Additionally, often coupled with the notion of direction is the concept of step size; in continuous problems, algorithms have a larger freedom in choosing the step size, whereas for combinatorial problems only integer increment or decrement are allowed.

3.5.2 Convex Versus Non-convex Optimization

The first classification introduced, between combinatorial and continuous optimization problems, is very general, and gives us an idea about the broad range of problems we may encounter when dealing with optimization. Nevertheless, the demarcation line between "easy" and "hard" optimization problems can be found in the distinction between convex and nonconvex optimization.

An important class of optimization problems is characterized by problems with *convex* objective function and constraints, which means that satisfy the inequality:

$$f(\alpha x + \beta y) \leq \alpha f(x) + \beta f(y) \tag{3.3}$$

for all $x, y \in \mathbb{R}^n$, $\alpha, \beta \in [0, 1]$ with $\alpha + \beta = 1$, and $\alpha \geq 0, \beta \geq 0$.

Graphically, a convex function lies on or below the line segment connecting any two points of the graph. A function f is strictly convex if strict inequality holds in (3.3), for every α, $0 < \alpha < 1$, $x \neq y$. When the objective function or any of the constraints are not convex, the problem is called a *nonconvex* optimization problem.

Convex optimization problems present some properties that make them particularly interesting, both from a theoretical and practical perspective. First, when a local optimum exsists for a convex problem, it is also a global optimum for the problem; this represents a key advantage of convex optimization, since avoids the stagnation into local optimal solutions. Additionally, for each strictly convex function, if the function has a minimum, then the minimum is unique. From a theoretical point of view, the associated dual problem offers interesting interpretations in terms of the original problem, with important practical consequences such as the design of efficient or distributed methods for solving it.

In general, when dealing with optimization problems, it is useful to verify whether the problem is convex or not. Such verification can be addressed by checking the definition of convexity, or showing that the function is obtained from simple convex functions by operations that preserve convexity, e.g. affine transformation or nonnegative weighted sum.

In order to verify the convexity, we can also consider the sufficient conditions of convexity.

Definition 2 (First-Order Conditions). Suppose $f : X \rightarrow Y$ is differentiable (i.e., its gradient ∇f exists at each point $x \in X$, which is open). Then f is convex if and only if X is convex and $f(y) \geq f(x) + \nabla f(x)^T (y - x)$ for all $x, y \in X$.

Definition 3 (Second-Order Conditions). Let $f : X \rightarrow Y$ be twice differentiable (i.e., its Hessian or second derivative $\nabla^2 f$ exists at each point $x \in X$, which is open). Then f is convex if and only if $\nabla^2 f(x) \preceq 0$, for all $x \in X$, i.e., its Hessian is positive semidefinite on its domain.

Convex optimization plays also an important role in addressing non-convex problems [1]; for instance, by reformulating a non-convex as an approximate convex we can obtain a solution to be used as a starting point for a local optimization method, applied for solving the original non-convex problem. Another application of convex optimization to non-convex problems consists in using relaxation methods, where non-convex constraints are replaced by convex ones; in Lagrangian relaxation, by solving the Lagrangian dual problem we obtain a lower bound on the optimal value of the original non-convex problem. From a computational point of view, convex problems can be efficiently solved by interior point methods [22,25], making convex relaxations very useful in practice.

3.5.3 Linear Versus Nonlinear Programming

A special case of convex optimization problem is represented by *linear programming* (LP) problems [4], where the objective and constraint functions are linear, i.e., satisfy the following:

$$f_i(\alpha x + \beta y) = \alpha f_i(x) + \beta f_i(y) \tag{3.4}$$

$\forall x, y \in \mathbb{R}^n$, and $\alpha, \beta \in \mathbb{R}$.

By looking at Problems 3.3 and 3.4, it is clear that convexity is more general than linearity. If the objective function or any of the constraints is not a linear function of the variables, the model is *nonlinear* and belongs to the class of the *non-linear programming* (NLP) problems.

The general form of a NLP problem is given as follows:

$$\begin{aligned} \min \ & f(x) \\ & g_i(x) \leq 0, \ i = 1, \ldots, m \\ & h_j(x) = 0, \ j = 1, \ldots, k \end{aligned} \tag{3.5}$$

where $f : \mathbb{R}^n \to \mathbb{R}$, $g_i : \mathbb{R}^n \to \mathbb{R}$, $h_j : \mathbb{R}^n \to \mathbb{R}$. Therefore, we can consider the linear programming problem as a special case of Problem 3.5, where

$$f_i(x) = \sum_{j=1}^{n} c_j x_j \tag{3.6}$$

and

$$g_i(x) = \sum_{j=1}^{n} a_{ij} x_j \quad i = 1, \ldots, m. \tag{3.7}$$

By adding the following constraints, the nonnegativity restrictions on variables for a linear programming problem are taken into account:

$$g_{m+i}(x) = -x_i \leq 0 \quad i = 1, \ldots, n. \tag{3.8}$$

The standard form of linear programming enables us to concisely represent the linear programming instance using linear algebra; the problem can be rewritten as

$$\begin{aligned} \min \ & c^T x \\ & A(x) \leq b \\ & x \geq 0 \end{aligned} \tag{3.9}$$

where $f(x) = c^T x$ and

$$g(x) = \begin{bmatrix} Ax - b \\ -x \end{bmatrix}. \tag{3.10}$$

The general method for solving a linear-programming problem is the simplex algorithm, developed by Dantzig [4]; the basic idea consists in graphing the inequality constraints on the Cartesian plane to draw a shape of a polygon, called "feasibility region". This region includes all the points that satisfy all the constraints. If the objective function is evaluated at all the point of the feasible region, the optimal value (if there is one) will occur at a vertex of such region.

Besides the simplex method, several other efficient methods for solving linear programming problems have been developed, including the ellipsoid algorithm, and several interior point algorithms. The availability of efficient methods is important because many large-scale real-world optimization problems are linear programming problems, involving a great number of variables and constraints. Typical examples are the organization and allocation of resources, portfolio optimization, vehicle routing, Internet routing, production line balancing.

Unlike linear programming, nonlinear optimization problems may be convex or not, with consequences on the difficulty of solving them. Since non-convex problems can be very challenging to solve, the traditional methods involve a compromise between quality of the solution and computational efficiency: on the one hand, local optimization methods are computationally efficient, but they cannot guarantee the global optimality of the solution found; on the other side, global optimization methods are able to find the global optimum, but can be very expensive in terms of computational time and resources.

For this reason, many approaches consist in decomposing the problem in subproblems and exploit the convexity of such subproblems to solve the original nonconvex problem. One approach is the use of branch and bound techniques, where the subproblems are solved with convex or linear approximations: the basic idea is to partition the feasible set into convex sets, and find the lower/upper bound for each subproblem; when the bound indicates that a subset cannot contain an optimal solution, this subset is discarded. The method is repeated until an optimal solution is found.

The optimality of a solution for a nonlinear optimization problem can be proved by verifying the Karush–Kuhn–Tucker (KKT) conditions [17]. KKT conditions are necessary for an optimal solution of a nonlinear program; under convexity, the KKT conditions are also sufficient.

Let us consider a nonlinear optimization problem, with the objective function $f(x)$ and the constraint functions $g_i(x)$ and $h_j(x)$ continuously differentiable at a point x^*. Let x^* be a local minimum for (3.5) that satisfies some regularity conditions. Then there exists a vector λ_j, $j = 1, \ldots, l$ where l represents the number of equality constraints, and μ_i, $i = 1, \ldots, m$, where m is the number of inequality constraints, such that the following conditions hold:

Definition 4 (Stationarity).

$$\nabla f(x^*) + \sum_{i=1}^{m} \lambda_j \nabla h_i(x^*) + \sum_{j=1}^{l} \mu_i \nabla g_j(x^*) = 0 \qquad (3.11)$$

Definition 5 (Primary Feasibility (PF)).

$$\begin{aligned} h_j(x^*) &= 0 \ j = 1, \ldots, l \\ g_i(x^*) &\leq 0 \ i = 1, \ldots, m \end{aligned} \qquad (3.12)$$

The Primal Feasibility states that inequality and equality constraints on x^* must be met in order x^* to be an optimal solution for the problem.

Definition 6 (Dual Feasibility (DF)).

$$\mu_i \geq 0 \quad i = 1, \ldots, m \qquad (3.13)$$

The Dual Feasibility states that every element μ_i must be greater or equal to zero.

Definition 7 (Complementary Slackness).

$$\mu_i g_i(x^*) = 0 \quad i = 1, \ldots, m \qquad (3.14)$$

λ_j and μ_i are called KKT multipliers. In the particular case when $m = 0$, that is no inequality constraints are present, the KKT conditions turn into the Lagrange conditions, and the KKT multipliers are called Lagrange multipliers.

In conclusion, for any nonlinear optimization problem with differentiable objective and constraint functions, any optimal point must satisfy the KKT conditions.

3.5.3.1 Integer, Binary, and Mixed-Integer Programming

When some or all the variables of a linear programming problem are required to be integer, the problem is called an *integer linear programming* (ILP), or more simply, an *integer programming* (IP) problem [30]. As for linear programs, the constraints for an integer program form a polytope, but in contrast to linear programming, the feasible points are all the integer-valued points within the polytope, therefore the feasible region is not a convex set. Additionally, whereas in linear programming the optimal solution occurs at a vertex of the polytope, for integer programming it lies at an extreme point of the convex hull of all feasible integer-valued points. It follows that while linear programming problems can be solved efficiently in the worst case, integer programming problems are NP-hard. This indicates that the worst-case time of any algorithm increases exponentially with the size of the problem, unless P = NP.

A special case of IP is represented by *0-1 integer linear programming*, known also as *binary programming* (BP), where variables are required to be 0 or 1. When only some of the unknown variables are required to be integer, the problems is called *mixed integer programming* (MIP) problem. Both BP and MIP are NP-hard.

As for the case of non-convex optimization problems, relaxation techniques, branch-and-bound, cutting-plane algorithms can be applied to solve IP problems, along with heuristics and population-based evolutionary algorithms. Another strategy consists in using hybrid methods, that combine the best features of both exact and heuristic strategies. A description of these approaches is provided in Sect. 3.6.

3.6 Methods

The available techniques for solving optimization problems can be roughly classi-fied into three main categories: *exact*, *approximation*, and *heuristic* methods. While exact methods allow to find solutions with theoretical guarantees, their run-time increases dramatically with the instance size. Heuristic approaches, on the other side, sacrifice optimality guarantees to find solutions in a more efficient way, by spending only a "reasonable" amount of time to compute them. Unlike heuristics, where there is no knowledge of the quality of the solution returned, approximation methods provide solutions within a small constant factor of the optimal solution. The combination of these approaches defines the category of *hybrid methods*.

3.6.1 Exact Methods

Exact methods have the advantage of guaranteed quality of the solution but the computational demand of such methods can be exponential in the worst case, and, in practice, often only moderately sized instances can be solved to optimality. Exact methods include a wide variety of techniques, such as Branch-and-Bound[13], Branch-and-Cut [19], Branch-and-Price [2], and Bayesian Optimization [28]. Nevertheless, in general finding an optimal solution is an intractable problem; although many results of mathematical analysis assure the presence of global optimizers for many problems, it is common to face real-world applications where exact methods are not applicable, due to the roughness of the search landscape or the difficulty in approximating derivatives.

3.6.2 Approximation Methods

In many cases, the requirements for optimality are relaxed towards the concept of satisfactory solutions; in this context, a major role is played by approximated algorithms. Formally, a ρ-approximation algorithm for a maximization problem is defined as a heuristic that runs in polynomial time and always returns a feasible solution whose value is within a factor of ρ from the value of an optimal solution. The quality of an approximate solution can be defined in terms of distance from the optimum, which we want to be as small as possible; in this case, ρ is referred to as the performance guarantee. Performance guarantees may be absolute, meaning that the additive difference between the optimal value and the value found by the algorithm is bounded. However, it is more common to find relative performance guarantees, where the value found by the approximation algorithm is within a multiplicative factor of the optimal solution. For some problems, we can obtain extremely good approximation algorithms, that characterize the polynomial-time approximation scheme (PTAS), where the solution found is within a factor $1 + \epsilon$ of being optimal (in the case of a minimization problem).

3.6.3 Heuristics Algorithms

Heuristic methods are applied when it is not possible to exploit the mathematical structure of the problem, or the resources required in terms of CPU and memory are prohibitive. In general, heuristic methods require the iterative application of a sampling method and a selection criterion; first, the algorithm generates solutions belonging to the search space using a predefined strategy, e.g. random perturbation of the variables, and, successively, it selects one solution from the sampled ones to use in the next iteration.

Although heuristic methods come with no performance guarantee, they perform well in practice and are able to provide satisfactory solutions in reasonable amount of time. Moreover, heuristics can be also used to improve the performance of an optimizer, by identifying promising starting solutions or by locating a promising search space. Additionally, due to their relatively simple features, heuristics are easy to implement even for complex problems. Effective heuristics are characterized by reasonable computational efforts, such as reasonable storage requirements and computing times, that do not grow exponentially as the problem size increases. Another important feature is their robustness, i.e., the method should not be sensitive to changes in parameters and have good average performance. The most effective heuristic approaches comprise Simulated Annealing [14, 16], Genetic Algorithms [11, 12, 31], Evolutionary Strategies [18], Tabu Search [10].

Metaheuristic algorithms extend heuristic methods by adopting a trade-off between randomization and local search, through intensification and diversification mechanisms, or exploitation and exploration [9]. Diversification concerns the creation of solutions that are different, in order to explore the search space and avoid local optima; intensification focuses more intently on regions previously exploited and found to be good. One way to classify metaheuristics consists in distinguishing between population-based and trajectory-based methods. Genetic algorithms, Particle Swarm [15], Ant Colony Optimization [5] belong to the first class. On the other hand, Simulated Annealing, Tabu Search, Variable Neighborhood Search [21], Greedy Randomized Adaptive Search [7] are trajectory-based methods, since they use a single solution which traces a trajectory on the search space.

3.6.4 Hybrid Methods

Hybrid methods combine the advantages of both exact and approximation/heuristic approaches: they mainly aim at providing solutions with optimality guarantee in shorter time. Hybrid methods can be mainly classified into two categories: *collaborative combinations* and *integrative combinations* [29].

The first class defines the algorithms that exchange information but are implemented as separate methods: each method can be executed in sequence, in parallel, or can be intertwined. Such approaches have been broadly proposed in literature, for instance evolutionary algorithms have been combined with branch-and-bound techniques, ant colony optimization algorithms have been combined with approximation algorithms, such as local search, and heuristic strategies have been applied to provide a pruning of the search space and speed up the computation of optimal solutions.

When two or more methods are combined into integrative combinations, one technique is a subordinate embedded component of the other. This is the case where metaheuristics are used within an exact method to determine bounds and incumbent solutions in branch-and-bound approaches, or the case of heuristic column generation in branch-and-price algorithms. Another example of integrative combinations is the exploration of neighborhoods in local search heuristics by means of exact algorithms.

References

1. Androulakis, I., Maranas, C., Floudas, C.: αbb: A global optimization method for general constrained nonconvex problems. J. Glob. Optim. **7**(4), 337–363 (1995)
2. Barnhart, C., Johnson, E.L., Nemhauser, G.L., Savelsbergh, M.W., Vance, P.H.: Branch-and-price: column generation for solving huge integer programs. Oper. Res. **46**(3), 316–329 (1998)
3. Boyd, S., Vandenberghe, L.: Convex Optimization. Cambridge University Press, Cambridge (2004)
4. Dantzig, G.B.: Linear Programming and Extensions. Princeton University Press, Princeton (1998)
5. Dorigo, M., Stützle, T.: The ant colony optimization metaheuristic: Algorithms, applications, and advances. In: Handbook of Metaheuristics, pp. 250–285. Springer, New York (2003)
6. Du, D.Z., Pardalos, P.M.: Handbook of Combinatorial Optimization, vol. 3. Springer, New York (1998)
7. Feo, T.A., Resende, M.G.: Greedy randomized adaptive search procedures. J. Glob. Optim. **6**(2), 109–133 (1995)
8. Garey, M.R., Johnson, D.S.: Computers and Intractability; A Guide to the Theory of NP-Completeness. WH Freeman, San Francisco (1990)
9. Gendreau, M.: Handbook of Metaheuristics, vol. 146. Springer, New York (2010)
10. Glover, F., Laguna, M.: Tabu Search. Wiley, London (1993)
11. Goldberg, D.E.: Genetic Algorithms in Search, Optimization, and Machine Learning. Addison-Wesley, Reading (1989)
12. Hansen, N., Müller, S.D., Koumoutsakos, P.: Reducing the time complexity of the derandomized evolution strategy with covariance matrix adaptation (CMA-ES). Evol. Comput. **11**(1), 1–18 (2003)
13. Horst, R., Pardalos, P.M., Van Thoai, N.: Introduction to Global Optimization. Kluwer, Dordrecht (2000)
14. Ingber, L., Petraglia, A., Petraglia, M.R., Machado, M.A.S., et al.: Adaptive simulated annealing. In: Stochastic Global Optimization and Its Applications with Fuzzy Adaptive Simulated Annealing, pp. 33–62. Springer, New York (2012)
15. Kennedy, J., Eberhart, R.: Particle swarm optimization. In: Proceedings of IEEE International Conference on Neural Networks, vol. 4, pp. 1942–1948. IEEE, New York (1995)
16. Kirkpatrick, S., Gelatt, D.G. Jr., Vecchi, M.P.: Optimization by simulated annealing. Science **220**(4598), 671–680 (1983)

17. Kuhn, H.W., Tucker, A.W.: Nonlinear programming. In: Proceedings of the Second Berkeley Symposium on Mathematical Statistics and Probability, vol. 5 (1951)
18. Michalewicz, Z.: Evolution strategies and other methods. In: Genetic Algorithms + Data Structures = Evolution Programs, pp. 159–177. Springer, New York (1996)
19. Mitchell, J.E.: Branch-and-cut algorithms for combinatorial optimization problems. In: Handbook of Applied Optimization, pp. 65–77. Oxford, GB: Oxford University Press (2002)
20. Mitchell, J.E., Pardalos, P.M., Resende, M.G.: Interior point methods for combinatorial optimization. In: Handbook of Combinatorial Optimization, vol. 1, pp. 189–297. Kluwer Academic Publishers (1998)
21. Mladenović, N., Hansen, P.: Variable neighborhood search. Comput. Oper. Res. **24**(11), 1097–1100 (1997)
22. Nesterov, Y., Nemirovskii, A.S., Ye, Y.: Interior-point Polynomial Algorithms in Convex Programming, vol. 13. Studies in Applied Mathematics, Philadelphia (1994)
23. Neumaier, A.: Complete search in continuous global optimization and constraint satisfaction. Acta Numer. **13**(1), 271–369 (2004)
24. Papadimitriou, C.H., Steiglitz, K.: Combinatorial Optimization: Algorithms and Complexity. Dover, Mineola (1998)
25. Pardalos, P.M., Resende, M.G.: Interior point methods for global optimization. In: Interior Point Methods of Mathematical Programming. Citeseer (1996)
26. Pardalos, P.M., Resende, M.G.: Handbook of Applied Optimization, vol. 1. Oxford University Press, Oxford (2002)
27. Pardalos, P.M., Rosen, J.B.: Constrained Global Optimization: Algorithms and Applications. Springer, New York (1987)
28. Pelikan, M.: Bayesian optimization algorithm. In: Hierarchical Bayesian Optimization Algorithm, pp. 31–48. Springer, New York (2005)
29. Puchinger, J., Raidl, G.: Combining metaheuristics and exact algorithms in combinatorial optimization: a survey and classification. In: Artificial Intelligence and Knowledge Engineering Applications: A Bioinspired Approach, pp. 113–124 (2005)
30. Schrijver, A.: Theory of Linear and Integer Programming. Wiley, London (1998)
31. Storn, R., Price, K.: Differential evolution–a simple and efficient heuristic for global optimization over continuous spaces. J. Glob. Optim. **11**(4), 341–359 (1997)

Chapter 4
String Selection Problems

Abstract The increasing amount of genomic data and the ability to synthe-size artificial DNA constructs poses a series of challenging problems involving the identification and design of sequences with specific properties. We address the identification of such sequences; many of these problems present challenges both at biological and computational level. In this chapter, we introduce the main string selection problems and the theoretical and experimental results for the most important instances.

4.1 Introduction

String selection problems (SSP) are among the most challenging problems in com-putational biology; they address the identification of similarities and/or differences in biological sequences, which is an underlying task of many molecular biology and biomedical applications.

In general, this class of problems can be formulated in terms of combinatorial optimization problems. In this chapter, we give a detailed overview of some of the most important problems in this field: first, we introduce the class of median string selection problems, aiming at finding a string that minimizes the sum of distances from the other elements of the input set; successively, we study the family of problems addressing the identification of closest strings, including the closest string problem (CSP), the closest substring problem (CSSP), the approximate substring problem (CAS), and the close to most string problem (CMSP). We will analyze the class of problems concerning the recognition of differences among biological sequences, which comprises the farthest string problem (FSP), the farthest substring problem (FSSP), and the far from most string problem (FFMSP). In this context, the distinguishing string selection problem (DSSP) can be viewed as a combination of both the CSP and FSP, since it looks for a string that occurs in a set of sequences but, at the same time, is different from another set; while the d-mismatch problem generalizes the concept of closest string by considering aligned set of substrings.

E. Pappalardo et al., *Optimization Approaches for Solving String Selection Problems,* SpringerBriefs in Optimization, DOI 10.1007/978-1-4614-9053-1__4,
© Elisa Pappalardo, Panos M. Pardalos, Giovanni Stracquadanio 2013

In each section, we introduce the basic definitions and successively we describe the state-of-the-art approaches proposed in literature; finally, we provide the reader with an extensive analysis on the computational results presented and their impact on solving real-world applications.

4.2   The Set Median String and the Median String Problems

Many biological problems involve the identification of a sequence which minimizes the sum of distances from a specific set of sequences; this problem is largely known as the *median string problem* (MSP), and it central to many bioinformatics applications. For instance, computing a median string can be viewed as solving the multiple tree alignment problem with a star phylogeny tree [37, 39], which is a well-studied and challenging problem in bioinformatics (see Sect. 2.4.3).

Definition 8 (Median String Problem). Let U be the space of all strings over the alphabet Σ, and S be a set of n strings from U. The median string s of S is defined as follows

$$s = \arg\min_{p \in U} \sum_{q \in S} d(p, q)$$

where d is a distance function, such as the conventional or the weighted edit distance (see Sect. 2.1). The string s is called a *generalized median* of S.

When the solution has to be a sequence belonging to the input set, the problem is known as the *set median string problem* [42, 48, 60].

Definition 9 (Set Median String Problem). Let U be the space of all strings over the alphabet Σ, and S be a set of n strings from U. The set median string s of S is defined as follows:

$$s = \arg\min_{p \in S} \sum_{q \in S} d(p, q)$$

and s is called a *set median* of S.

The set median represents an approximation of the generalized median string, and the restriction to the input set makes the problem solvable in polynomial time. It has been proved that computing the median string is NP-complete when the alphabet Σ is unbounded [39], and NP-hard for alphabet of size 7 and using a conveniently weighted edit distance [77]. Additionally, the MSP has been demonstrated to be NP-complete and W[1]-hard with respect to the number of input strings, under the Levenshtein distance, even when Σ is binary [73, 74].

Practically, W[1]-hardness can be interpreted as the fact that a parameterized problem with parameter k is unlikely to have an algorithm with running time $f(k)n^{\mathcal{O}(1)}$, that is the parameterized problem is fixed parameter intractable.

4.2.1 Methods

Due to their computational complexity, several approximate algorithms have been proposed in literature for both set median and median search problems. When the distance function is a metric, one approach for solving the Set Median String Problem is to avoid distance computations [44], being quadratic in space and time [70].

An approximate algorithm, which uses an estimate of the sum of the distances, instead of computing the exact sum of the distances of each string to all the other strings in S, is presented in [70]. The algorithm relies on two steps: in the first one, a subset of n_r strings from S is selected, and the sum of distances is computed. In the second phase, the n_t strings having lowest partial sum are considered, and the sum of distances to the remaining strings are computed. The string that minimizes the full sum is then chosen as the approximate median. This algorithm has complexity $\mathcal{O}(n_r + n_t|S|)$, where S is the input set of strings; in practical situations, it runs in linear time [70].

A method that implements an iterative refinement procedure for set median computation is proposed in [60], running in $\mathcal{O}(m^3 n|\Sigma|)$, where m is the length of the strings, n the number of strings in S and Σ the alphabet. The method computes an initial approximate solution and iteratively tries to improve it by means of systematic perturbations, that are substitutions, insertions, or deletions. Due to its computational cost, this method has not found practical applications.

A first exact algorithm for the computation of generalized median string was proposed in [49], although it has exponential time complexity. Nevertheless, the knowledge of the specific domain gained by such approach is used in [57] to reduce the computation time, by considering the case when the sequences in S are quite similar. This leads to a complexity of $\mathcal{O}(mk^n)$, where m is the length of the sequences, n the number of strings in S, and k is a small constant used as an upper bound on the distance of the generalized median string. Though such complexity remains exponential, k is usually much smaller than m, making it suitable in practice.

In order to deal with the NP-hardness of generalized median string computation, a plethora of approximate and heuristic methods has been presented in literature. A greedy algorithm has been proposed in [11], which has time complexity of $\mathcal{O}(m^2 n|\Sigma|)$ when the Levenshtein distance is used as a metric, and space complexity of $\mathcal{O}(mn|\Sigma|)$. This method is improved in [50], where a new termination criterion and a new selection mechanism for breaking ties are introduced. Such modifications lead to the same time complexity, but slightly improve the space complexity to $\mathcal{O}(mn)$.

A new paradigm is introduced in [43], where strings are embedded into a vector space, and the median string is computed in such space, simplifying the computation of the distances. In [41], a dynamic input set is taken into account, where strings can be added once the computation of the generalized median has already started. To compute the new median string for S^{t+1}, the dynamic approach considers only the generalized median string for S^t and the new data item, but not the individual member of S^t [42]. This is done by exploiting a fundamental fact in real space: given that the generalized median of a set of points $S^t = \{p_1, p_2, \ldots, p_t\}$ is the well-known mean $\overline{p}^t = 1/t \sum_{i=1}^{t} p_i$, when a new point p_{t+1} is added to the initial set of points, the new set $S^t \cup \{p_{t+1}\}$ has generalized mean equal to $\overline{p}^{t+1} = 1/(t+1) \sum_{i=1}^{t+1} p_i = t/(t+1)\overline{p}^t + 1/(t+1)p_{t+1}$. This case is generalized to the domain of strings, avoiding the need to perform expensive operations, such as computing the median from scratch every time a new string is added to the current set of strings, thus making the algorithm suitable to a dynamic environment.

4.2.2   Computational Results

The methods presented in literature for MSPs have been tested on both synthetic and real data. Real data consist of a set of handwritten digits, extracted from the UNIPEN database [38], where each digit can be viewed as a sequence of 2D points $d = (x_1, y_1), \ldots, (x_m, y_m)$ on the xy-plane. Those sequences are transformed into strings by first resampling the data points such that the distance between any consecutive pair of points has a constant value Δ; d is transformed into a sequence $d' = (\overline{x}_1, \overline{y}_1), \ldots, (\overline{x}_m, \overline{y}_m)$, where $|(\overline{x}_{i+1}, \overline{y}_{i+1}) - (\overline{x}_i, \overline{y}_i)| = \Delta$, for $i = 1, \ldots, m-1$. At this point, a string $s = a_1 a_2 \ldots a_{m-1}$ is generated from d', where a_i is the vector from $(\overline{x}_i, \overline{y}_i)$ to $(\overline{x}_{i+1}, \overline{y}_{i+1})$ [41]. In [41], Δ has been fixed to 7, so that digit strings have between 40 and 100 points; 10 samples of digit 1, 2 and 3, and 98 samples of digit 6 are considered. The cost of insertions, deletions, and substitutions is fixed to $c(a \rightarrow \epsilon) = c(\epsilon \rightarrow a) = |a| = \Delta$, and $c(a, b) = |a - b|$, respectively, where ϵ denotes the empty string. The minimum cost of a substitution is equal to zero, which happens when $a = b$, and its maximum cost is equal to 2Δ, that occurs when a and b are parallel and have opposite direction. The methods proposed in [11] and [41] have been tested on such dataset, and the latter approach, which considers a dynamic context, performs better, while the greedy method [11] performs good on short strings, but encounters some difficulties in dealing with longer strings [42]. Various synthetic datasets have been tested for string median problems, ranging from strings based on the German alphabet of 59 letters and 10 digits [43], to natural forenames from the beginning of the Finnish calendar [48], to Spanish words [60]. Unfortunately, the lack of a standard testbed prevents a robust and comprehensive comparison among the methods proposed in literature.

4.3   The Closest and the Farthest String Problems

The *CSP*, also known as *Consensus String Problem*, defines the task of finding a pattern that, with some errors, occurs in a specific set of strings. In particular, it consists in finding a string with minimum Hamming distance from the strings of a given finite input set.

Definition 10 (Closest String Problem). Let $S = \{s^1, s^2, \ldots, s^n\}$ be a finite set of n strings of length m, over an alphabet Σ.

The CSP for S is to find a string t of length m over Σ, that minimizes the Hamming distance d_c such that, for any $s \in S$,

$$d_H(t, s) \leq d_c.$$

Let us consider the following example: given four strings on a binary alphabet $\Sigma = \{0, 1\}$, $S = \{001000, 111000, 011011, 010101\}$, the string $t = 011001$ is a *closest string* for S; in fact, $\max_{i=1}^{4} d_H(s^i, t) = \max\{2, 2, 1, 2\} = 2$.

Conversely, the *FSP* defines the problem of finding a pattern which does not occur in a set of strings.

Definition 11 (Farthest String Problem). Let $S = \{s^1, s^2, \ldots, s^n\}$ be a finite set of n strings of length m, over an alphabet Σ. The FSP for S is to find a string t of length m over Σ, that maximizes the Hamming distance d_f such that, for any $s \in S$,

$$d_H(t, s) \geq d_f.$$

The FSP can be considered as the complementary of the CSP, thus it can be solved by using similar techniques to that employed for the CSP, although these problems have been proved to be NP-hard; in particular, Frances and Litman have proved the NP-hardness of the CSP for binary codes, by considering an equivalent problem in coding theory [27]. Then, Lanctot et al. have shown that even in the case of alphabets with more than two characters, both the CSP and the FSP are NP-hard [51]; in order to prove its NP-hardness, the FSP has been reduced to the strong 3-SAT problem, thus the CSP has been reduced to the FSP. It follows that for such problems no polynomial-time solution is possible, unless "P = NP".

4.3.1   Methods

The first *integer-programming* (IP) formulation for the CSP has been proposed, independently, in [6] and [53], and in [51] for the FSP.

In [65], three IP formulations are presented for the problem along with a heuristic; the CSP is first reduced to an integer-programming problem, then a branch-and-bound algorithm is used to solve it. Specifically, an instance of the CSP consists of a finite set $S = \{s^1, s^2, \ldots, s^n\}$ of strings such that $|s^i| = m$. By s^i_j, we denote the jth position of the string s^i, for $1 \leq i \leq n$ and $1 \leq j \leq m$. The goal of the problem is to find a string $t \in \Sigma^m$ such that $\max_i d_H(s^i, t)$, $1 \leq i \leq n$, is minimized, where d_H represents the Hamming distance. Instead of working directly on strings, an injective transformation π is applied, which maps each character c into an integer $\pi(c)$, and any string $s = (s_1, s_2, \ldots, s_m)$ into a sequence $\pi(s) = (x_1, x_2, \ldots, x_m) \in \mathbb{Z}$ of the same length, such that $x_k = \pi(s_k)$, for $k = 1, \ldots, m$.

As an example, given the alphabet $\Sigma = \{a, \ldots, z\}$, let us consider the canonical transformation $\pi(a) = 1, \pi(b) = 2, \ldots, \pi(z) = 26$. Then, the string $s = differ$ is mapped into $\pi(s) = (4, 9, 6, 6, 5, 18) \in \mathbb{Z}^6$. Given a set $S = \{s^1, \ldots, s^n\}$ of strings of the same length m, V_k represents the set of the kth characters of the strings s_i, that is $V_k = \{s^1_k, s^2_k, \ldots, s^n_k\}$, for $i = 1, \ldots, n$ and $k = 1, \ldots, m$.

Additionally, the binary variables $v_{j,k}$ are defined, which characterize a candidate solution t as follows:

$$v_{j,k} = \begin{cases} 1 & \text{if } t_k = j \\ 0 & \text{otherwise} \end{cases} \tag{4.1}$$

for $j \in V_k, k = 1, \ldots, m$.

Finally, the integer-programming formulation for the CSP is defined as follows [6, 53, 65]:

$$\min \ d_H \tag{4.2}$$

subject to

$$\sum_{j \in V_k} v_{j,k} = 1 \qquad k = 1, \ldots, m \tag{4.3}$$

$$m - \sum_{j=1}^{m} v_{s^i_j, j} \leq d_H \qquad i = 1, \ldots, n \tag{4.4}$$

$$v_{j,k} \in \{0, 1\} \qquad j \in V_k, \quad k = 1, \ldots, m \tag{4.5}$$

$$d_H \in \mathbb{Z}_+ \tag{4.6}$$

The objective function (4.2) minimizes the Hamming distance d_H, whereas (4.3) and (4.5) guarantee that exactly one character in V_k is selected. Inequalities (4.4) say that if a character in a string s^i is not chosen for a solution t, then such character will contribute to increase the Hamming distance between t and s^i. Finally, constraints (4.6) force the distance d_H to assume a nonnegative value. The above formulation (4.2)–(4.6) is the most effective from a computational point of view, since it involves fewer variables and constraints [13, 65]. Nevertheless, the approach proposed in [65] is not always efficient and has exponential-time complexity, and the branch-and-bound technique leads easily to memory explosion.

In order to speed up the solution search, a parallel version of the heuristics presented in [65, 67] is developed in [32]. In particular, the difference between the method proposed in [65] and [67] lies on the local search procedure, which in the latter uses several processors on a parallel machine.

A new formulation for the CSP is proposed in [13], although experimental results prove that it is not as effective as the one proposed in [6, 53] and [65].

Like the CSP, the FSP can be formulated as an integer program and similar techniques can be used to solve the resulting optimization problem. In particular, the mathematical formulation of the FSP is similar to the one used for the CSP, with only a change in the optimization objective (4.2), which becomes:

$$\max \ d_H \tag{4.7}$$

and the inequality sign in the constraint, which is redefined as follows

$$m - \sum_{j=1}^{m} v_{s_j^i, j} \geq d_H \quad i = 1, \ldots, n. \tag{4.8}$$

Though the FSP finds many applications in biology, the majority of attention has been focused on the CSP. Nevertheless, since CSP and FSP represent two sides of the same optimization problem, it is straightforward to extend CSP approaches to the FSP.

Due to the limited efficiency of exact methods in solving large instances of the problems, a large number of non-exact approaches have been proposed in literature. In general, as described in Sect. 3.5, we can distinguish between two classes of non-exact methods for solving optimization problems: approximation algorithms and heuristics. Approximation algorithms guarantee a bound on the optimal solution, whereas heuristic strategies do not.

Several approximation algorithms have been developed for the CSP problem: two $(4/3 + \epsilon)$-polynomial-time approximation algorithms have been developed in [28, 51]. In [51], also a PTAS for the FSP has been proposed, which is based on the randomized rounding of the relaxed solution found by solving the ILP formulation (4.7)–(4.8). In [16], the previous results are extended to take into account short strings with length m in the range of $\mathcal{O}(\log n)$, for unbounded alphabet size, so that the number of possible characters $|\Sigma|$ is considered as a parameter.

To overcome the NP-hardness of the problem, a fixed-parameter algorithm with running time $\mathcal{O}(nm + nd^{d+1})$ has been developed for the CSP in [33, 34], where d is a parameter representing the maximum Hamming distance allowed. The underlying idea of this approach is to study the parameterized complexity of the problem, under the assumption that either d or the number of input strings n are small. Plainly, for large values of d, such approach becomes prohibitive.

Based on the previous approaches, new polynomial-time approximation algorithms for the CSP having approximation ratio $1 + \epsilon$, for any small ϵ, have been

developed [53]. A further improvement is presented in [59], where an algorithm
that finds the optimal solution for CSP with running time $\mathscr{O}(mn + nd(16|\Sigma|)^d)$ is
proposed. The complexity results are improved in [83], where a $\mathscr{O}(mn + nd(|\Sigma - 1|^d 2^{3.25d})$-time fixed parameter algorithm and a $\mathscr{O}(mn + nd 2^{3.25d_b})$-time fixed
parameter algorithm for the CSP and FSP are presented, respectively.

Recently, a polynomial-time approximation algorithm, called three-string
approach, has been presented [12]. This method starts by selecting three of the input
strings which are used to guess a portion of the output center string, outperforming
the previous approaches, which, in most cases, are based on the two-string approach.

In [14], an LP-based approach is presented, which consists in solving the
continuous relaxation of the ILP formulation of CSP. Specifically, three algorithms
have been developed and their results are compared with the solutions provided in
[56], where an exact algorithm for the case of three strings with alphabet size equal
to two is presented; however, this case is only of theoretical interest. A new approach
has been introduced in [47], where the CSP is addressed as a constraint satisfaction
problem. Despite the innovation proposed by this method, the results are claimed
just for short instances of the problem that do not represent a real application. In
[1], an exact method having running time $\mathscr{O}(m^2)$ for only five binary strings, has
been presented. Nevertheless, this case is of theoretical importance only.

Generally speaking, the running time of the approximation algorithms in the case
of large set of instances makes them not practical for real applications.

Since heuristics are usually able to provide good solutions in reasonable amount
of time, many heuristic approaches have been proposed for the CSP. Genetic
Algorithms (GA) [7, 31, 40] are among the most known metaheuristic methods;
they mimic the natural evolutionary process by evolving a population of solutions. Genetic algorithms are based on natural selection and sexual reproduction
processes; the first mechanism determines which members of a population survive
and are able to reproduce, the second one assures genetic recombination among
individuals of the same population. The principle of selection is based on a function,
called fitness, that measures "how good an individual is": individuals with better
fitness have higher probability to reproduce and survive. Due to their computational
power and easy implementation, several genetic algorithms have been developed for
the CSP.

The first GA has been proposed in [63]; such paradigm has been successively
adopted in [45, 54]. For the CSP, the most natural representation for an individual
is a string over the defined alphabet; it follows that the population is a collection of
strings. The fitness function used to evaluate the quality of a solution is the objective
function of the problem, and a lower fitness function represents a better string, that
is a solution minimizing the maximum Hamming distance with the strings of the
input set S. In [18], a genetic algorithm which uses a new distance metric, the rank
distance, has been introduced. Informally, the rank distance between two strings
measures the gap between the positions of a character in the two given strings,
and then sums up these values. Intuitively, the rank distance gives us the total non-
alignment score between two sequences. The edit Levenshtein distance is taken into
account in [29], where a further genetic algorithm is presented. A *memetic algorithm*

(MA) has been proposed in [4] for both CSP and FSP, which integrates a local search strategy within a GA. A fundamental concept in memetic algorithms is the neighborhood-based local search; in [4], the neighborhood of a solution s is defined as the set of strings having Hamming distance equal to one from s. At the end of each generation of the genetic algorithm, the local search is performed, which is intended to improve the current solution by exploring the neighborhood structure.

Besides GAs, several others heuristic approaches have been successfully applied to solve SSP. A *simulated annealing* approach (SA) has been presented in [46] and [54], and a hybrid algorithm has been implemented in [55], which combines both the Genetic and Simulated Annealing approaches, though limited only to binary alphabets. Simulated Annealing is a generalization of Monte Carlo methods, originally proposed by Metropolis and Ulam [68, 69] as a means of finding the equilibrium configuration of a collection of atoms at a given temperature. The basic idea of SA was taken from an analogy with the annealing process used in metallurgy, a technique involving heating and controlled cooling of a material to increase the size of its crystals and reduce their defects. Methods based on SA apply a probabilistic mechanism to escape local minima: the underlying idea is to accept, under certain conditions, not only transitions that improve the objective function value, but also transitions that do not. The probability of accepting worsening steps varies during the search phase, and it slowly decreases to zero. In the original Metropolis scheme, an initial state (or solution) is chosen, having energy E and temperature T. Keeping T constant, such initial configuration is perturbed, and the energy change ΔE is computed. If ΔE is negative, the new solution is always accepted, otherwise it is accepted with a probability given by the Boltzmann factor $e^{-(\Delta E/T)}$. This process is repeated L times for the current temperature, then the temperature is decremented and the entire process is repeated until a frozen state is reached at $T = 0$. Due to such characteristics, methods based on the SA may accept not only transitions that lead to better solutions, but also transitions that lead to pejorative ones, though with probabilities which tend to zero: at the beginning of the search, when temperatures are high, the algorithm behaves as a random search and therefore bad solutions can be accepted; whereas for lower values of T, solutions are located in promising regions of the search space. Like genetic algorithms for the CSP, a solution in SA is a sequence of characters, and solutions minimizing the Hamming distance value have higher probability of being accepted for the next iteration of the algorithm.

Another metaheuristic approach applied to solve the CSP consists of the *ant-colony optimization algorithm* (ACO) [20,21] in [23]. ACO metaheuristic represents a transposition of real ants behavior to artificial intelligence and has been inspired by the observation of real ant colonies, where the behavior of each single ant is directed to the survival of the whole colony. In particular, in his analysis Dorigo pointed out the foraging behavior of ants: when a new food source is found, ants search for the shortest and easiest way to return to their nest. While walking from the food source to the nest, and vice versa, ants deposit on the ground a chemical substance called pheromone. Ants can smell pheromones and, when choosing their way, they select with higher probability paths marked by stronger pheromone concentrations. It has been proved that pheromone trails make shortest paths to emerge over other paths, due to the fact that pheromone density tends to be higher on such paths. In view of all these facts, artificial ants are modeled on the behavior of real ants. In short, the ACO algorithm for CSP works as follows: a set of asynchronous and concurrent agents, a colony of ants, is initialized. Each ant builds a solution, by choosing, for each position m, a character from the alphabet. Ants choose their way probabilistically, using a probability depending on the value of the local pheromone trail, associated with each character. When each ant has built its own solution, pheromone trails are updated; the larger is the pheromone trail for a character, the higher will be the probability that this character will be chosen in the next iteration. After additional pheromone trail on the best string has been released, the evaporation procedure is applied, which consists in decrementing pheromone values, as pheromone density tends to increase on shortest paths. This process leads to the emergence of the best closest string among all the solutions created by artificial ants.

In [75], a greedy randomized adaptive search procedure (GRASP, for short) metaheuristic has been presented, along with a new probabilistic heuristic function, used to discriminate among candidate solutions having the same objective value. GRASP consists of two main processes, the construction phase and the local search phase. The construction phase builds an initial solution, to be optimized during the local search. After building an initial solution, a Greedy Randomized solution is constructed at each iteration of the algorithm. In order to select the elements to add to the current solution, a restricted candidate list (RCL) is used which is based on the heuristic function. The local search phase receives the candidate solution built by the construction phase and RCL and tries to improve it. A local search strategy is presented also in [56], which consists in iteratively replacing a character with a new one that does not make the solution worse.

A hybrid approach is presented in a recent work [80], where a heuristic strategy based on a Lagrangian relaxation is combined with a Tabu Search method for the CSP. Specifically, the Lagrangian relaxation provides a tight lower bound and an approximate solution for a mixed-integer programming formulation of the problem. After the approximate solution is found, a tabu search method tries to improve it. Tabu search methods implement a local search mechanism that relies on specialized memory structures in order to avoid stagnation into local minima: this is done by keeping track of the previously found solutions, so that if a potential solution has been already visited within a certain short-term period, it is marked as "taboo."

4.3.2   Computational Results

The effectiveness of the approaches introduced in Sect. 4.3.1 has been assessed by extensive testing on real and simulated data. The real dataset consists of instances from the McClure dataset [64], characterized by an alphabet size $|\Sigma| = 20$; these strings represent a set of protein sequences frequently used to test string-comparison algorithms. McClure data consist of six instances: for three of them, $m = 141$

and $n = \{6, 10, 12\}$; two datasets have length $m = 98$ and $n = \{10, 12\}$; and the last set is characterized by $m = 100$ and $n = 6$. The methods proposed in [32, 56, 65] have been tested and compared on this dataset; it turns out that solving the IP formulation proposed in [65] always leads to the optimal solution. In fact, though solving the exact ILP results in exponential runtime behavior, in practice integer programming techniques are effective in solving the CSP for the McClure dataset, due to the moderate-size of its instances.

Instances for the simulated data have been randomly generated in many different works, by varying the number of strings n, their length m, and the alphabet size $|\Sigma|$. Three alphabet sizes have been studied, $|\Sigma| = \{2, 4, 20\}$, which consider, respectively, binary strings ($\Sigma = \{0, 1\}$), nucleotide bases ($\Sigma = \{A, C, G, T\}$), and amino-acid sequences. Usually n varies in $\{10, 20, 30, 40, 50\}$, and m in $\{100, 200, \ldots, 900, 1000, 2000, \ldots, 5000\}$. As for real datasets, although a large number of heuristic methods have been developed for the CSP, solving the ILP formulation gives the best results in terms of solution quality. Though it is not possible to compare the running times of all the proposed methods, since they depend on the computing platform and on the features of the algorithm implemented, experimental results show that as either m or n increase, running times become dramatically higher. For such reason, heuristic are preferable when the problem size increases. Additionally, it seems that the running time of the ILP approach is strongly dependent on the structure of the sequences of the input set, and specifically it is affected by the ratio between the length of the string m, and the number of mismatches among the strings of each input set. Experimental results show that an increasing length and a thereby increasing ratio can significantly decrease the running time of the problem [13, 34]. With respect to the alphabet size, in general it turns out that small alphabet sizes are harder to solve, in terms of computational time required to find an optimal solution. The same consideration can be assessed for the FSP [67].

4.4   The Closest and the Farthest Substring Problems

The CSP and the FSP can be viewed as a special case of two more general problems, respectively, the *CSSP* and the *FSSP*, which are more difficult to solve. Intuitively, CSSP and FSSP search for a short string that is enriched or not in each string in a given set of sequences.

Formally, the CSSP and the FSSP can be defined as follows.

Definition 12 (Closest Substring Problem). Let $S = \{s^1, s^2, \ldots, s^n\}$ be a finite set of n strings of length at least m, over an alphabet Σ. The CSSP for S is to find a string t of length m over Σ, that minimizes the Hamming distance d_{cs} for every string $s^i \in S, i = 1, \ldots, n$, such that

$$d_H(t, u) \leq d_{cs} \tag{4.9}$$

where u is *some* length m substring of s^i.

As an example, suppose $S = \{$"$ACGTCA$", "$TTAC$", "$CCGC$"$\}$ and $m = 4$. The string $t = $ "$CTGC$" satisfies $d_H(t, u) \leq d_{cs}$ for some substring u of $s^i \in S$ with $d_{cs} = 2$ [66].

Definition 13 (Farthest Substring Problem). Let $S = \{s^1, s^2, \ldots, s^n\}$ be a finite set of n strings of length at least m, over the alphabet Σ. The FSSP for S is to find a string t of length m over Σ, maximizing the Hamming distance d_{fs}, such that for *every* string $s^i \in S$, $i = 1, \ldots, n$, and every length m substring u of s^i,

$$d_H(t, u) \geq d_{fs} \tag{4.10}$$

For instance, suppose $S = \{$"$ACGTCA$", "$TTAC$", "$CCGC$"$\}$ and $m = 4$. The string $t = $ "$AACG$" satisfies $d_H(t, u) \geq d_{fs}$ for every substring u of $s^i \in S$ with $d_{fs} = 3$ [66].

In terms of parameterized complexity,[1] the main result for the CSSP is its fixed parameter intractability. This is expressed by showing that the CSSP belongs to the complexity class of W[1]-hard problems, with respect to the number k of input strings, even if the alphabet is binary [24].

4.4.1 Methods

As a consequence of the W[1]-hardness of CSSP, the problem is unlikely to achieve running time $f(k)n^{\mathcal{O}(1)}$, for any function f, i.e. exponential only in the number of strings k [24]. From a practical point of view, when $k \geq 3$ the problem is expected to be intractable. Such results have been improved in [61], where the CSSP parameterized by the distance d is proved to be W[1]-hard even for binary alphabet. However, this does not completely exclude the existence of feasible PTASs for the problem.

Although several PTASs have been designed for the CSSP [53, 58], they cannot be applied in practice due to their very high computational complexity. The first ratio-2 polynomial-time approximation algorithm for the CSSP has been studied in [51], and the results presented in this work have been improved in [52], where the first nontrivial algorithm with approximation ratio $2 - \frac{2}{2|\Sigma|+1}$ is presented. Starting from these results, a PTAS with running time of $\mathcal{O}(mn^{O(\epsilon^{-2} \log 1/\epsilon)})$, and one with further improved time complexity $\mathcal{O}(mn^{O(\epsilon^{-2})})$ are proposed, respectively, in [2] and [59].

The complexity results provided in [24] are further extended in [81, 82], where it is showed that the CSSP problem on unbounded alphabet has no PTAS of running time $f(1/\epsilon)|x|^{o(1/\epsilon)}$ for any function f, where $|x|$ is the size of input instance.

---

[1]Informally, the goal of parameterized complexity is to study how the different parameters of the input instance affect the running time of the algorithm.

According to such lower bound, it follows that the problem has no practical PTAS, even when the error bound $\epsilon > 0$ is small. The same results hold in the case of bounded alphabets: even in this case, the CSSP has no PTAS of running time $f(1/\epsilon)|x|^{o(\log(1/\epsilon))}$ for any function f [61]. This result does not exclude that the problem has a PTAS of running time $f(1/\epsilon)|x|^{O(\log(1/\epsilon))}$. In particular, in [61] two algorithms are presented; the first has running time $|\Sigma|^{d(\log d+2)}|x|^{O\log d}$, where $|x|$ is the total length of input strings. This algorithm might be efficient for small values of d, since only the logarithm of the parameter appears in the exponent of $|x|$. The second algorithm runs in $|\Sigma|^d \cdot 2^{kd} \cdot d^{O(d \log \log k)} \cdot |x|^{O(\log \log k)}$, where k is the number of input strings.

The complementary maximization version of the CSSP, the Max Closest Substring, is studied in [22, 78], where it is proved that the problem cannot be approximated in polynomial time with ratio better than $(\log k)/4$, where k is the number of strings, unless "P = NP".

As for the CSP and FSP, the CSSP and FSSP can be modeled as integer programming problems [66]. Specifically, let us define the input set $S = \{s^1, \ldots, s^n\}$ of n strings s^i, and let V_k be the set of characters appearing at the kth position of strings in S. $s^{i,k}$ denotes the kth substring of length m of string s^i, that is we start with the m first characters of s^i, and then move one character to right at a time.

Let us introduce the variables $x_{j,k}$ and $y_{i,k}$ as follows:

$$x_{j,k} = \begin{cases} 1 & \text{if character } j \text{ is used at } k\text{th position in a solution} \\ 0 & \text{otherwise} \end{cases} \tag{4.11}$$

$$y_{i,k} = \begin{cases} 1 & \text{if length-}m \text{ substring } s^{i,k} \text{ of } s^i \in S \text{ satisfies } d_H(x, s^{i,k}) \leq d_{cs} \\ 0 & \text{otherwise} \end{cases}$$

$$\tag{4.12}$$

Therefore, the integer-programming formulation for the CSSP is defined as :

$$\min \ d_{cs} \tag{4.13}$$

subject to

$$\sum_{j \in V} x_{j,k} = 1 \qquad k = 1, \ldots, m \tag{4.14}$$

$$d_{cs} + \sum_{j=1}^{m} x_{s_j^{i,k},j} \geq m y_{i,k} \qquad i = 1, \ldots, n \quad k = 1, \ldots, |s^i| - m + 1 \tag{4.15}$$

$$\sum_{k=1}^{|s^i|-m+1} y_{i,k} \geq 1 \qquad i = 1, \ldots, n \tag{4.16}$$

$$x_{j,k} \in \{0,1\} \qquad\qquad j \in V, \quad k = 1,\ldots,m \qquad\qquad (4.17)$$

$$y_{i,k} \in \{0,1\} \quad i = 1,\ldots,n, \quad k = 1,\ldots,|s^i| - m + 1 \quad (4.18)$$

$$d_{cs} \in \mathbb{Z}_+ \qquad\qquad\qquad\qquad\qquad\qquad\qquad\qquad (4.19)$$

Equalities (4.14) guarantee that only one character in V is chosen for each position; inequalities (4.15) and (4.16) ensure that $d_H(x,s^{i,k}) \leq d_{cs}$ for at least one substring $s^{i,k}$ of s^i having length m. Constraints (4.17)–(4.18) state binary variables, and (4.19) forces d_{cs} to assume a nonnegative integer value.

The FSSP can be modeled similarly to the CSSP. Specifically, FSSP is modeled as an IP problem as follows:

$$\max\ d_{fs} \qquad\qquad\qquad\qquad\qquad\qquad (4.20)$$

subject to

$$\sum_{j \in V} x_{j,k} = 1 \qquad\qquad k = 1,\ldots,m \qquad\qquad (4.21)$$

$$m - \sum_{j=1}^{m} x_{s_j^{i,k},j} \geq d_{fs} \quad i = 1,\ldots,n \quad k = 1,\ldots,|s^i| - m + 1 \quad (4.22)$$

$$x_{j,k} \in \{0,1\} \qquad\qquad j \in V, \quad k = 1,\ldots,m \qquad\qquad (4.23)$$

$$d_{fs} \in \mathbb{Z}_+ \qquad\qquad\qquad\qquad\qquad\qquad (4.24)$$

Here equalities (4.21) guarantee that, for each position of the solution, only one character in V is chosen; inequalities (4.22) force x to satisfy $d_H(x,s^{i,k}) \geq d_{fs}$, for every length-$m$ substring $s^{i,k}$ of $s^i \in S$; constraints (4.23) force the variables to assume binary values, and (4.24) forces the distance d_{fs} to assume a nonnegative integer value.

To solve the IP formulations for CSSP and FSSP, a branch-and-bound method has been proposed in [66]. Nevertheless, such approach becomes prohibitive for real-world instances of the problems. For such reason, some heuristic methods have been developed.

An evolutionary algorithm for CSSP is presented in [62]; as for the CSP, the most natural representation for a candidate solution is a string over the alphabet Σ. Therefore the population consists of collection of strings from Σ. An Ant-Colony strategy is developed in [5], and the experimental results have been compared with the genetic algorithm presented in [62]. A genetic algorithm that uses a fitness function based on rank distance for CSSP has been developed in [19]. As for the CSP, a GRASP metaheuristic for FSSP is presented in [76]. Here path-relinking is adopted to keep track of the previous iterations, namely, to incorporate memory into GRASP. This strategy can be viewed as an intensification mechanism which uses existing solutions to influence the current iteration; shortly, the current solution is combined with a solution from an elite set, using path-relinking operator, which creates a path on the solution space. Such mechanism seeks to isolate attributes that occur in high quality solutions, in order to favor the choices of the assignments that lead to better solutions.

A different approach for the CSSP has been proposed in [10], where a parallel GPU exact algorithm is presented. Though the parallel implementation of the algorithm, the nature of the exact algorithm makes it impracticable for practical use.

4.5 The Close to Most String Problem

The presence of strings which are significantly different from the other strings in the input set dramatically affects the ability of finding a minimal solution for the CSP and FSP. In some cases, such strings can be classified as outliers and can be excluded from the search, by seeking for similarity in a restricted portion of the input set. This leads to the definition of the CMSP, also known as the *max close string problem* (MCSP), where given a parameter d and a set S of strings having the same length, the objective is to find a string s which maximizes the number of non-outliers within Hamming distance d of s [9].

Definition 14 (Close to Most String Problem). Given n strings $S = \{s^1, \ldots, s^n\}$ of length m over an alphabet Σ, and $d \in \mathbb{Z}_+$, the objective is to find a string s of length m such that the number of strings s^i in S satisfying $d(s, s^i) \leq d$ is maximized.

Differently from what shown in [58], this problem is proved to not have a PTAS in [9], unless $ZPP^2 = NP$, by using a reduction from the Max-2-SAT Problem. In [8], a fixed parameter algorithm where d and the number of outliers k are the parameters is provided, for both bounded and unbounded alphabets.

As for the CSP and CSSP, the CMSP can be modeled as an Integer Programming Problem [66].

Let us introduce $x_{j,k}$ and $y_{i,k}$ as follows:

$$x_{j,k} = \begin{cases} 1 & \text{if character } j \text{ is used at } k\text{-th position in a solution} \\ 0 & \text{otherwise} \end{cases} \qquad (4.25)$$

$$y_{i,k} = \begin{cases} 1 & \text{if length-}m \text{ string } s^i \in S \text{ satisfies } d_H(x, s^i) \leq d_{cms} \\ 0 & \text{otherwise} \end{cases} \qquad (4.26)$$

Then, the integer-programming formulation for the CMSP can be modeled as :

---

[2]ZPP is the *Zero-error Probabilistic Polynomial Time* complexity class. It is defined as the class of languages recognized by probabilistic Turing machine with polynomial bounded average run time and zero error probability [30].

$$\max \sum_{i=1}^{n} y_i \tag{4.27}$$

subject to

$$\sum_{j \in V_k} x_{j,k} = 1 \qquad k = 1, \ldots, m \tag{4.28}$$

$$d_{cms} + \sum_{j=1}^{m} x_{s_j^i, j} \geq m y_i \qquad i = 1, \ldots, n \tag{4.29}$$

$$x_{j,k} \in \{0, 1\} \qquad j \in V_k, \quad k = 1, \ldots, m \tag{4.30}$$

$$y_{i,k} \in \{0, 1\} \qquad i = 1, \ldots, n \tag{4.31}$$

$$d_{cms}, m \in \mathbb{Z}_+ \tag{4.32}$$

Equalities (4.28) guarantee that only one character in each V_k is chosen for each position; inequalities (4.29) say that if a character in a string s^i is not in a solution x, then it contributes to increase the Hamming distance from x to s^i. Constraints (4.30)–(4.31) state binary variables, and (4.32) forces d_{cs} and m to assume a nonnegative integer value.

A branch-and-bound algorithm is used in [66] to solve the proposed formulation, and it is tested on both randomly generated instances on a four characters alphabet, and on the McClure dataset [64]. Though, as expected, the linear relaxation leads to a more efficient computation, no comparison can be made with other approaches, since there is no literature on the experimental results for the CMSP.

A slightly different problem considers the positions in a string as outliers, rather than the strings themselves, and it is known as the *Most Strings with Few Bad Columns Problem* [9]. Informally, the objective here is to maximize the number of strings with at most k columns having entries not-all-equal. The problem is proved to be NP-hard and APX-hard,[3] which means that if $P \neq NP$, there is not a PTAS for the problem.

4.6 The Far From Most String Problem

A problem that is closely related to the FSP concerns the search of a string which is far from most of the input strings in S. Such problem, known as the FFMSP, shares a significant similarity with the CMSP introduced in the previous section. Below we formally define the problem [67].

---

[3]APX is defined as the class of all NP-optimization problems P such that, for some $r \geq 1$, there exists a polynomial time r-approximate algorithm for P [3].

Definition 15 (Far From Most String Problem). Given a threshold t and the input set S, a string s must be found maximizing the variable x such that $d_{\text{ffms}}(x, s^i) \geq t$, for $s^i \in P \subseteq S$, and $|P| = x$.

Despite the similarity with the FSP, the FFMSP is much harder to approximate [25, 67].

Due to its hardness, polynomial-time algorithms do not provide any constant guarantee of approximation; for such reason, several heuristic approaches have been proposed in literature. A simple heuristic is developed in [67], which favors the selection of characters that increase the number of differences between the strings in S and the current solution. Such current solution is then refined by a local search method, which implements a two-exchange procedure.

A GRASP is presented in [25], which leads to an improve of 10–15 % with respect to the previously proposed heuristic [67]. In [26], this method is compared to a genetic algorithm, which always produces better quality solutions.

A new metaheuristic method, combining a constructive and a local search phase, is proposed in [71]. Specifically, the first phase builds several candidate solutions by implementing a constructive beam search algorithm: a heuristic function evaluates candidate solutions based on their likelihood to lead to better solutions by performing as few changes as possible. Once several candidate solutions are obtained, a local search procedure seeks to improve them. The best solution found among all the feasible ones is returned as the final solution to the problem. In order to evaluate its performance, the proposed method has been compared to the GRASP algorithm presented in [25]. Computational results show that the newly proposed method performs at least as good as GRASP on every test instance, and outperforms it in 45 cases out of 54.

The same heuristic strategy presented in [71] is incorporated within a GRASP in [72]. The main difference with the GRASP method presented in [25] lies in the heuristic evaluation function adopted in the local search phase. This leads to a significant improvement of the quality of the solutions found.

In order to measure how the new heuristic function affects the search landscape, particularly on reducing the number of local maximum points, the average number of uphill moves in the local search phases for both methods are reported in [72]; in fact, an uphill move implies that the point is not a local, or a global, optimum. Therefore, the higher the number of such points is, the lower is the number of local or global maximum points. On average, the new proposed algorithm leads to an increase of 96.32 and 95.45 % of uphill moves for random and real instances, respectively. This represents a good result, since it is well known how the existence of many local optima causes a degradation of the performance of metaheuristics strategies.

4.7   The Distinguishing String and Substring Selection Problems

So far, we have considered problems where the objective is to find a string that minimizes or maximizes the distance for an input set of strings, but we did not take into account the case when these two objectives have to be attained at the same time.

The DSSP and the *distinguishing substring selection problem* (DSSSP) on the other hand aim at finding a string which, concurrently, minimizes the distance with the strings of an input set, and maximizes the distance with another set of strings. Informally, the objective is to find a solution string which is close to a set of "good strings," and far from a set of "bad strings."

Formally, the DSSP and DSSSP problems can be defined as follows [51, 83].

Definition 16 (Distinguishing String Selection Problem). Given two sets of strings S_c and S_f, all of length m, and two positive integers d_c and d_f, with $d_f \geq m - d_c$, the DSSP is to find a string t of length m such that the Hamming distance $d(t, s_c) \leq d_c$, and for any string $s_c \in S_c$, and $d(t, s_f) \geq d_f$, for any $s_f \in S_f$.

Definition 17 (Distinguishing Substring Selection Problem). Given two sets of strings S_c and S_f, all of length at least m, and two positive integers d_c and d_f, $d_c \leq d_f$, the DSSSP is to find a string t such that for each string s_c in S_c, there exist a length m substring y_c of s_c such that $d(t, y_c) \leq d_c$, and for any substring y_f of s_f in S_f, $d(t, y_f) \geq d_f$.

The DSSP is proved to be NP-hard in [51], where a polynomial time factor two-approximation algorithm is given for such problem. In [33], the DSSP has been proved to be solvable in time $\mathcal{O}((|S_c| + |S_f|)m \cdot \max(d_c + 1, (d_f + 1)|\Sigma|)^{d_c})$, by adapting a recursive algorithm based on the bounded search tree paradigm developed for the CSP. This method implements a branch of the solution space, and in case of constant alphabet size and distance parameters, runs in linear time.

A PTAS for the DSSSP is presented in [15, 17], where for any constant $\epsilon > 0$, the algorithm computes a center string t of length m such that, for every $s_c \in S_c$, there is a length m substring y_c of s_c with $d(t, y_c) \leq (1 + \epsilon)d_c$, and for every substring y_f of $s_f \in S_f$, of length m, $d(t, y_f) \geq (1 - \epsilon)d_f$, if a solution exists.

Since the CSP (see Sect. 4.3) is a special case of the DSSSP, and it has been proved to be W[1]-hard, it follows that DSSSP is also W[1]-hard even for binary alphabets with respect to the number of input strings and the distance parameter; so the problem is fixed-parameter intractable, as proved in [35, 36].

Such result implies that no efficient PTAS (EPTAS) are available for the problem, unless FPT[4] = W[1], therefore even the PTAS presented in [15] appears to be impractical. Despite such theoretical conclusion, it seems the border between

---

[4]FPT denotes the class of fixed-parameter tractable problems, which are problems that can be solved in time $f(k)|x|^{\mathcal{O}(1)}$ for some computable function f.

fixed-parameter tractability and intractability lies between alphabet sizes two and three: this result is proved by studying a modified version of the DSSSP, where the problem is restricted to a binary alphabet, a "dual distance parameterizations" is employed for good strings and such parameter $d'_f = m - d_f$ is required to be optimal [35, 36]. This modified version is showed to be fixed-parameter tractable with respect to the parameter d'_f, while extending the alphabet size from two to three makes the problem combinatorially much more difficult. According to such result, a fixed-parameter algorithm for the binary DSSP is designed in [83], having running time $\mathcal{O}(m|S_c| + |S_c|d\,2^{3.25d_c})$.

4.8 The d-Mismatch Problem

Another generalization of the CSP is the d-mismatch Problem, which searches for a center string in the case of aligned substrings of an input set.

Formally, let $s^i_{p,L}$ denote the length-L substring of a given string s^i starting at position p. Then, given a set of strings $S = \{s^1, s^2, \ldots, s^n\}$ of length m, and integers L and d, the d-mismatch problem is to find a string t of length L and a position p with $1 \le p \le m - L + 1$, such that $d_H(t, s^i_{p,L}) \le d$ for all $i = 1, \ldots, n$ [33, 34]. When $m = L$, the problem is equivalent to the CSP. It follows that the d-mismatch problem is also NP-hard.

A linear-time algorithm for the 1-mismatch problem has been presented in [79], and an algorithm which runs in linear time for constant d is presented in [33, 34]. Specifically, these latter works prove that d-mismatch is solvable in time $\mathcal{O}(nL + (m - L)nd \cdot d^d)$, which is $\mathcal{O}(mn)$ for constant d. Though these methods might seem impractical for high values of d, in some biological application small distance parameters d are not unusual, such as in primer design [34], making the presented approach suitable in these cases.

4.9 Conclusions

In this work, we presented a compendium of the most challenging String Selection problems, by first introducing the mathematical and biological background, and then discussing the state-of-the-art approaches presented in literature.

A large number of problems in molecular biology can be formulated as combinatorial optimization problems, and the applications of operation research models and methods are rapidly increasing in this field. Although many theoretical analyses exist for these problems, there is still a significant gap between biological and computational results, suggesting that the difficulty of solving many of the studied problems does not depend only on their computational hardness but also on their biological structure and properties [65].

From a computational point of view, still many topics remain open for further research; since SSP are computationally intractable, efficient and effective algorithms that guarantee an optimal solution within a reasonable amount of time are unlikely to be proposed. Many questions remain to be answered from a theoretical perspective: among these, the W[1]-completeness of CSSP [34], the existence of an EPTAS for CSP [9], as well the existence of a PTAS for the MSP [73].

From a biological point of view, the fast improvement in DNA synthesis technologies will provide new challenging problems in this area, especially in the area of genome engineering and sequencing. In silico approaches play a major role in such contexts, and the necessity of effective optimization methods leads to a new focus on these subjects.

References

1. Amir, A., Paryenty, H., Roditty, L.: Configurations and minority in the string consensus problem. In: String Processing and Information Retrieval, pp. 42–53. Springer, Berlin (2012)
2. Andoni, A., Indyk, P., Patrascu, M.: On the optimality of the dimensionality reduction method. In: 47th Annual IEEE Symposium on Foundations of Computer Science, 2006 (FOCS'06), pp. 449–458. IEEE, New York (2006)
3. Ausiello, G.: Complexity and approximation: Combinatorial optimization problems and their approximability properties. Springer, Berlin (1999)
4. Babaie, M., Mousavi, S.: A memetic algorithm for closest string problem and farthest string problem. In: 18th Iranian Conference on Electrical Engineering (ICEE), pp. 570–575. IEEE, New York (2010)
5. Bahredar, F., Javadi, H., Moghadam, R., Erfani, H., Navidi, H.: A meta heuristic solution for closest substring problem using ant colony system. Adv. Stud. Biol. 2(4), 179–189 (2010)
6. Ben-Dor, A., Lancia, G., Ravi, R., Perone, J.: Banishing bias from consensus sequences. In: Combinatorial Pattern Matching, pp. 247–261. Springer, Berlin (1997)
7. Booker, L., Goldberg, D., Holland, J.: Classifier systems and genetic algorithms. In: Machine Learning: Paradigms and Methods Table of Contents, pp. 235–282 (1990)
8. Boucher, C., Ma, B.: Closest string with outliers. BMC bioinformatics, 12(Suppl 1), S55 (2011)
9. Boucher, C., Landau, G.M., Levy, A., Pritchard, D., Weimann, O.: On approximating string selection problems with outliers. In: Proceedings of the 23rd Annual Conference on Combinatorial Pattern Matching, pp. 427–438. Springer, Berlin (2012)
10. Calhoun, J., Graham, J., Jiang, H.: On using a graphics processing unit to solve the closest substring problem. In: International Conference on Parallel and Distributed Processing Techniques and Applications (PDPTA) (2011)
11. Casacuberta, F., de Antonio, M.: A greedy algorithm for computing approximate median strings. In: Proceedings of Spanish Symposium on Pattern Recognition and Image Analysis, pp. 193–198. AERFAI (1997)
12. Chen, Z.Z., Ma, B., Wang, L.: A three-string approach to the closest string problem. J. Comput. Syst. Sci., 78(1), 164–178 (2012)
13. Chimani, M., Woste, M., Böcker, S.: A closer look at the closest string and closest substring problem. In: Proceedings of the 13th Workshop on Algorithm Engineering and Experiments (ALENEX), pp. 13–24 (2011)
14. Della Croce, F., Salassa, F.: Improved lp-based algorithms for the closest string problem. Comput. Oper. Res. 39(3), 746–749 (2012)
15. Deng, X., Li, G., Li, Z., Ma, B., Wang, L.: A PTAS for distinguishing (sub)string selection. In: Automata, Languages and Programming, pp. 788–788 (2002)

16. Deng, X., Li, G., Wang, L.: Center and distinguisher for strings with unbounded alphabet. J. Comb. Optim. **6**(4), 383–400 (2002)
17. Deng, X., Li, G., Li, Z., Ma, B., Wang, L.: Genetic design of drugs without side-effects. SIAM J. Comput. **32**(4), 1073–1090 (2003)
18. Dinu, L., Ionescu, R.: A genetic approximation of closest string via rank distance. In: 13th International Symposium on Symbolic and Numeric Algorithms for Scientific Computing (SYNASC), pp. 207–214. IEEE, New York (2011)
19. Dinu, L., Ionescu, R.: An efficient rank based approach for closest string and closest substring. PloS One **7**(6), e37576 (2012)
20. Dorigo, M.: Optimization, learning and natural algorithms. Ph.D. thesis, Dipartimento di Elettronica, Politecnico di Milano (1992)
21. Dorigo, M., Caro, G., Gambardella, L.: Ant algorithms for discrete optimization. Artif. Life **5**(2), 137–172 (1999)
22. Evans, P., Smith, A.: Complexity of approximating closest substring problems. In: Fundamentals of Computation Theory, pp. 13–47. Springer, Berlin (2003)
23. Faro, S., Pappalardo, E.: Ant-CSP: An ant colony optimization algorithm for the closest string problem. In: SOFSEM 2010: Theory and Practice of Computer Science, pp. 370–381. Springer Berlin Heidelberg (2010)
24. Fellows, M., Gramm, J., Niedermeier, R.: On the parameterized intractability of closest substring and related problems. In: STACS 2002, pp. 262–273. Springer Berlin Heidelberg (2002)
25. Festa, P.: On some optimization problems in molecular biology. Math. Biosci. **207**(2), 219–234 (2007)
26. Festa, P., Pardalos, P.M.: Efficient solutions for the far from most string problem. Ann. Oper. Res. **196**(1), 663–682 (2012)
27. Frances, M., Litman, A.: On covering problems of codes. Theor. Comput. Syst. **30**(2), 113–119 (1997)
28. Gąsieniec, L., Jansson, J., Lingas, A.: Efficient approximation algorithms for the Hamming center problem. In: Proceedings of the Tenth Annual ACM-SIAM Symposium on Discrete Algorithms: Society for Industrial and Applied Mathematics, pp. 905–906 (1999)
29. Gilkerson, J., Jaromczyk, J.: The genetic algorithm scheme for consensus sequences. In: IEEE Congress on Evolutionary Computation, 2007 (CEC 2007), pp. 3870–3878. IEEE, New York (2007)
30. Gill, J.: Computational complexity of probabilistic turing machines. SIAM J. Comput. **6**(4), 675–695 (1977)
31. Goldberg, D., Holland, J.: Genetic algorithms and machine learning. Mach. Learn. **3**(2), 95–99 (1988)
32. Gomes, F., Meneses, C., Pardalos, P., Viana, G.: A parallel multistart algorithm for the closest string problem. Comput. Oper. Res. **35**(11), 3636–3643 (2008)
33. Gramm, J., Niedermeier, R., Rossmanith, P.: Exact solutions for closest string and related problems. Algorithms and Computation, pp. 441–453. Springer Berlin Heidelberg (2001)
34. Gramm, J., Niedermeier, R., Rossmanith, P.: Fixed-parameter algorithms for closest string and 743 related problems. Algorithmica **37**(1), 25–Ű42 (2003)
35. Gramm, J., Guo, J., Niedermeier, R.: On exact and approximation algorithms for distinguishing substring selection. In: Proceedings of Fundamentals of Computation Theory: 14th International Symposium (FCT 2003), Malmö, 12–15 August 2003, vol. 14, p. 195. Springer, Berlin (2003)
36. Gramm, J., Guo, J., Niedermeier, R.: Parameterized intractability of distinguishing substring selection. Theor. Comput. Syst. **39**(4), 545–560 (2006)
37. Gusfield, D.: Algorithms on Strings, Trees and Sequences: Computer Science and Computational Biology. Cambridge University Press, Cambridge (1997)
38. Guyon, I., Schomaker, L., Plamondon, R., Liberman, M., Janet, S.: UNIPEN project of on-line data exchange and recognizer benchmarks. In: Proceedings of the 12th IAPR International Conference on Pattern Recognition, vol. 2-Conference B: Computer Vision & Image Processing, vol. 2, pp. 29–33. IEEE, New York (1994)

39. de la Higuera, C., Casacuberta, F.: Topology of strings: median string is NP-complete. Theor. Comput. Sci. **230**(1), 39–48 (2000)
40. Holland, J.: Adaptation in Natural and Artificial Systems. MIT, Cambridge (1992)
41. Jiang, X., Abegglen, K., Bunke, H., Csirik, J.: Dynamic computation of generalised median strings. Pattern Anal. Appl. **6**(3), 185–193 (2003)
42. Jiang, X., Bunke, H., Csirik, J.: Median strings: a review. In: Data Mining in Time Series Databases, pp. 173–192 (2004)
43. Jiang, X., Wentker, J., Ferrer, M.: Generalized median string computation by means of string embedding in vector spaces. Pattern Recognit. Lett. **33**(7), 842–852 (2012)
44. Juan, A., Vidal, E.: Fast median search in metric spaces. In: Advances in Pattern Recognition, pp. 905–912. Springer Berlin Heidelberg (1998)
45. Julstrom, B.: A data-based coding of candidate strings in the closest string problem. In: Proceedings of the 11th Annual Conference Companion on Genetic and Evolutionary Computation Conference: Late Breaking Papers, pp. 2053–2058. Association for Computing Machinery (2009)
46. Keith, J., Adams, P., Bryant, D., Kroese, D., Mitchelson, K., Cochran, D., Lala, G.: A simulated annealing algorithm for finding consensus sequences. Bioinformatics **18**(11), 1494–1499 (2002)
47. Kelsey, T., Kotthoff, L.: The exact closest string problem as a constraint satisfaction problem. Arxiv preprint arXiv:1005.0089 (2010)
48. Kohonen, T.: Median strings. Pattern Recognit. Lett. **3**(5), 309–313 (1985)
49. Kruskal, J.B.: An overview of sequence comparison: time warps, string edits, and macro-molecules. SIAM Rev. **25**(2), 201–237 (1983)
50. Kruzslicz, F.: Improved greedy algorithm for computing approximate median strings. Acta Cybern. **14**(2), 331–340 (1999)
51. Lanctot, J.K., Li, M., Ma, B., Wang, S., Zhang, L.: Distinguishing string selection problems. In: Proceedings of the tenth annual ACM-SIAM symposium on Discrete algorithms, pp. 633–642. Society for Industrial and Applied Mathematics (1999)
52. Li, M., Ma, B., Wang, L.: Finding similar regions in many strings. In: Proceedings of the Thirty-first Annual ACM Symposium on Theory of computing, pp. 473–482. Association for Computing Machinery (1999)
53. Li, M., Ma, B., Wang, L.: On the closest string and substring problems. J. ACM **49**(2), 157–171 (2002)
54. Liu, X., He, H., Sýkora, O.: Parallel genetic algorithm and parallel simulated annealing algorithm for the closest string problem. In: Advanced Data Mining and Applications, pp. 591–597. Springer Berlin Heidelberg (2005)
55. Liu, X., Holger, M., Hao, Z., Wu, G.: A compounded genetic and simulated annealing algo-rithm for the closest string problem. In: The 2nd International Conference on Bioinformatics and Biomedical Engineering, 2008 (ICBBE 2008), pp. 702–705. IEEE, New York (2008)
56. Liu, X., Liu, S., Hao, Z., Mauch, H.: Exact algorithm and heuristic for the closest string problem. Comput. & Oper. Res., **38**(11), 1513–1520 (2011)
57. Lopresti, D., Zhou, J.: Using consensus sequence voting to correct OCR errors. Comput. Vis. Image Underst. **67**(1), 39–47 (1997)
58. Ma, B.: A polynomial time approximation scheme for the closest substring problem. In: Combinatorial Pattern Matching, pp. 99–107. Springer, Berlin (2000)
59. Ma, B., Sun, X.: More efficient algorithms for closest string and substring problems. In: Research in Computational Molecular Biology, pp. 396–409. Springer, Berlin (2008)
60. Martínez-Hinarejos, C.D., Juan, A., Casacuberta, F.: Use of median string for classification. In: Proceedings of 15th International Conference on Pattern Recognition, vol. 2, pp. 903–906. IEEE, New York (2000)
61. Marx, D.: Closest substring problems with small distances. SIAM J. Comput. **38**(4), 1382–1410 (2008)

62. Mauch, H.: Closest substring problem–results from an evolutionary algorithm. In: Neural Information Processing, pp. 205–211. Springer, Berlin (2004)
63. Mauch, H., Melzer, M., Hu, J.: Genetic algorithm approach for the closest string problem. In: Proceedings of the 2003 IEEE Bioinformatics Conference 2003 (CSB 2003), pp. 560–561 (2003)
64. McClure, M., Vasi, T., Fitch, W.: Comparative analysis of multiple protein-sequence alignment methods. Mol. Biol. Evol. **11**(4), 571 (1994)
65. Meneses, C., Lu, Z., Oliveira, C., Pardalos, P., et al.: Optimal solutions for the closest-string problem via integer programming. INFORMS J. Comput. **16**(4), 419–429 (2004)
66. Meneses, C., Pardalos, P., Resende, M., Vazacopoulos, A.: Modeling and solving string selection problems. In: Second International Symposium on Mathematical and Computational Biology, pp. 54–64 (2005)
67. Meneses, C., Oliveira, C., Pardalos, P.: Optimization techniques for string selection and comparison problems in genomics. IEEE Eng. Med. Biol. Mag. **24**(3), 81–87 (2005)
68. Metropolis, N., Ulam, S.: The Monte Carlo method. J. Am. Stat. Assoc. **44**(247), 335–341 (1949)
69. Metropolis, N., Rosenbluth, A., Rosenbluth, M., Teller, A., Teller, E.: Perspective on "Equation of state calculations by fast computing machines". J. Chem. Phys. **21**, 1087–1092 (1953)
70. Micó, L., Oncina, J.: An approximate median search algorithm in non-metric spaces. Pattern Recognit. Lett. **22**(10), 1145–1151 (2001)
71. Mousavi, S.R.: A hybridization of constructive beam search with local search for far from most strings problem. Int. J. Comput. Math. Sci. **v4**(i7), 340–348 (2010)
72. Mousavi, S.R., Babaie, M., Montazerian, M.: An improved heuristic for the far from most strings problem. J. Heuristics **18**(2), 239–262 (2012)
73. Nicolas, F., Rivals, E.: Complexities of the centre and median string problems. In: Combinatorial Pattern Matching, pp. 315–327. Springer, Berlin (2003)
74. Nicolas, F., Rivals, E.: Hardness results for the center and median string problems under the weighted and unweighted edit distances. J. Discrete Algorithms **3**(2), 390–415 (2005)
75. Mousavi, S.R., Nasr Esfahani, N.: A GRASP algorithm for the closest string problem using a probability-based heuristic. Comput. & Oper. Res., **39**(2), 238–248 (2012)
76. Silva, R.M.A., Baleeiro, G., Pires, D., Resende, M., Festa, P., Valentim, F.: Grasp with path-relinking for the farthest substring problem. Technical Report, AT&T Labs Research (2008)
77. Sim, J.S., Park, K.: The consensus string problem for a metric is NP-complete. J. Discrete Algorithms **1**(1), 111–117 (2003)
78. Smith, A.: Common approximate substrings. Ph.D. thesis, Citeseer (2004)
79. Stojanovic, N., Berman, P., Gumucio, D., Hardison, R., Miller, W.: A linear-time algorithm for the 1-mismatch problem. In: Algorithms and Data Structures, pp. 126–135. Springer Berlin Heidelberg (1997)
80. Tanaka, S.: A heuristic algorithm based on Lagrangian relaxation for the closest string problem. Comput. & Oper. Res., **39**(3), 709–717 (2012)
81. Wang, J., Huang, M., Chen., J.: A lower bound on approximation algorithms for the closest substring problem. In: Combinatorial Optimization and Applications, pp. 291–300. Springer Berlin Heidelberg (2007)
82. Wang, J., Chen, J., Huang, M.: An improved lower bound on approximation algorithms for the closest substring problem. Inf. Process. Lett. **107**(1), 24–28 (2008)
83. Wang, L., Zhu, B.: Efficient algorithms for the closest string and distinguishing string selection problems. In: Frontiers in Algorithmics, pp. 261–270. Springer Berlin Heidelberg (2009)